LIBERTY AND THE ECOLOGICAL CRISIS

This book examines the concept of liberty in relation to civilization's ability to live within ecological limits.

Freedom, in all its renditions – choice, thought, action – has become inextricably linked to our understanding of what it means to be modern citizens. And yet, it is our relatively unbounded freedom that has resulted in so much ecological devastation. Liberty has piggy-backed on transformations in human–nature relationships that characterize the Anthropocene: increasing extraction of resources, industrialization, technological development, ecological destruction, and mass production linked to global consumerism. This volume provides a deeply critical examination of the concept of liberty as it relates to environmental politics and ethics in the long view. Contributions explore this entanglement of freedom and the ecological crisis, as well as investigate alternative modernities and more ecologically benign ways of living on Earth. The overarching framework for this collection is that liberty and agency need to be rethought before these strongly held ideals of our age are forced out. On a finite planet, our choices will become limited if we hope to survive the climatic transitions set in motion by uncontrolled consumption of resources and energy over the past 150 years. This volume suggests concrete political and philosophical approaches and governance strategies for learning how to flourish in new ways within the ecological constraints of the planet.

Mapping out new ways forward for long-term ecological well-being, this book is essential reading for students and scholars of ecology, environmental ethics, politics, and sociology, and for the wider audience interested in the human–Earth relationship and global sustainability.

Christopher J. Orr is a PhD candidate as part of the Economics for the Anthropocene project in the Department of Natural Resource Sciences at McGill University, Canada.

Kaitlin Kish is a Postdoctoral Fellow for the Economics for the Anthropocene project at McGill University and lecturer at the University of British Columbia's Haida Gwaii Institute, Canada.

Bruce Jennings is Adjunct Associate Professor in the Department of Health Policy and the Center for Biomedical Ethics and Society at the Vanderbilt University School of Medicine. He is Senior Fellow at the Center for Humans and Nature in Chicago, and Senior Advisor and Fellow at The Hastings Center in New York.

ROUTLEDGE EXPLORATIONS IN ENVIRONMENTAL STUDIES

LIBERTY AND THE ECOLOGICAL CRISIS

Freedom on a Finite Planet

Edited by Christopher J. Orr, Kaitlin Kish, and Bruce Jennings

Routledge
Taylor & Francis Group

LONDON AND NEW YORK

earthscan
from Routledge

First published 2020
by Routledge
2 Park Square, Milton Park, Abingdon, Oxon OX14 4RN

and by Routledge
52 Vanderbilt Avenue, New York, NY 10017

Routledge is an imprint of the Taylor & Francis Group, an informa business

British Library Cataloguing-in-Publication Data
A catalogue record for this book is available from the British Library

Library of Congress Cataloging-in-Publication Data
A catalog record has been requested for this book

ISBN: 978-0-367-33933-3 (hbk)
ISBN: 978-0-367-34677-5 (pbk)
ISBN: 978-0-429-32710-0 (ebk)

Typeset in Bembo
by Deanta Global Publishing Services, Chennai, India

CONTENTS

ILLUSTRATIONS

Figures

Table

ACKNOWLEDGEMENTS

This collection was developed over several years through a series of presentations and workshops as part of the Economics for the Anthropocene project at McGill University in Montreal, Canada. Thank you to all the participants and authors for contributing to these events and working diligently with the editors to refine their chapters and develop this collection.

Financial support was provided by a Social Science and Humanities Research Council of Canada Partnership Grant, which funded a symposium that brought all the authors together to develop the ideas and resulted in this collection. Additional funding was provided by the Waterloo Institute for Complexity and Innovation at the University of Waterloo, the Economics for the Anthropocene research project, and the Department of Natural Resource Sciences at McGill University.

A special thank you to Dr. Peter Brown, Dina Spigelski, Joanna Mastalerek, and Eliza Zahirovic for all their hard work, support, and encouragement.

CONTRIBUTORS

Anna Beresford is a PhD candidate in the Social and Ecological Sustainability program at the University of Waterloo. She is researching how the cultural spaces of traditional music and dance contribute to local community resilience and economic sustainability. Working from an Aristotelian tradition, she is also examining the philosophical foundations behind place-making, social capital formation, sustainable development, and patterns.

Peter G. Brown is a Professor at McGill University. His work centres on the deterioration of Earth's life support capacity, and the thought systems that facilitate and legitimate this decline. He is an author, co-author, and editor of many books. He is actively involved in conservation efforts in the US and Canada. His property in Quebec contains healthy examples of several species at risk, and in Maryland he protects a pristine river. He is a member of the Society of Friends (Quakers) and the Club of Rome. In 2017 he received the Herman Daly Award for advancing the discipline of ecological economics.

Peter F. Cannavò is Professor of Government at Hamilton College. His work focuses on the political theory of place, the relationship between civic republicanism and environmentalism, and the normative dimensions of climate change. He is author of *The Working Landscape: Founding, Preservation, and the Politics of Place* (The MIT Press, 2007), and co-editor (with Joseph H. Lane, Jr.) of *Engaging Nature: Environmentalism and the Political Theory Canon* (The MIT Press, 2014). He has published in *Political Theory, Environmental Politics, Environmental Values,* and elsewhere. He is currently working on *To the Thousandth Generation: The Green Civic Republican Tradition in America.*

Martin Hensher is Associate Professor of Health Systems Financing and Organisation in Deakin University's Institute for Health Transformation, based in Melbourne, Australia. He has spent much of his career as a senior public servant working on health and health care policy in the United Kingdom, South Africa, and Australia. His current research is focused on examining how best to prepare public health policy and health care systems for the emerging ecological, economic, and social challenges of the Anthropocene era.

Bruce Jennings is Adjunct Associate Professor in the Department of Health Policy and the Center for Biomedical Ethics and Society at the Vanderbilt University School of Medicine. He is Senior Fellow at the Center for Humans and Nature in Chicago, and Senior Advisor and Fellow at The Hastings Center in New York. He has worked in the areas of bioethics, environmental ethics, and ecological political theory related to biotechnology and climate change. His most recent book is *Ecological Governance: Toward a New Social Contract with the Earth* (2106).

Aaron Karp is an activist focused on generating a discussion within the climate justice movement about the need for degrowth and deep, authentic democracy. He is writing a book about why our ecological crises demand economic and cultural transformation, not just an energy transition, and how the movement can lay the groundwork for these changes. In 2012 he co-founded the fossil fuel divestment campaign at UMass Amherst, where he studied neuroscience and Spanish. He writes about movement strategy at freedomsurvival.org.

Kaitlin Kish is a Postdoctoral Fellow for the Economics for the Anthropocene project at McGill University and lecturer at the University of British Columbia's Haida Gwaii Institute. Kaitlin's background is in systems thinking, ecological economics, and complexity science, which she applies to her current research on exploring radical and disruptive political economies and possible pathways to alternative futures – particularly related to new forms of production and manufacturing in localized economies. She is also working on disseminating and helping to formulate a wide-reaching research agenda for Ecological Economics through two special issues on a research agenda for Ecological Economics in the Journal of Sustainability and the Journal for Ecological Economics and contributions in other related publications. Kaitlin has a PhD in Social and Ecological Sustainability from the University of Waterloo.

Amy R. McCready is Associate Professor of Political Science at Bucknell University who has worked on early modern British thought. She is leaving academia to care for farm and woodlands in Pennsylvania.

Steven J. Mock (PhD, London School of Economics) is Research Director of the Ideological Conflict Project at the Balsillie School for International Affairs,

University of Waterloo, Canada. His research interests focus on methods for modelling the myths, symbols and rituals associated with the constructs of national and other forms of political-cultural identity, further to understanding the impact of these constructs on conflict and conflict resolution. He is the author of *Symbols of Defeat in the Construction of National Identity* (Cambridge University Press, 2012).

Jeffery L. Nicholas (PhD philosophy, University of Kentucky) is an Associate Professor at Providence College and an international scholar on ethics and politics. He serves as research associate for the Center for Aristotelian Studies in Ethics and Politics at London Metropolitan University and at the Center for Aristotle and Critical Theory at Mykolas Reomeris University in Lithuania. Dr. Nicholas is co-founder of and executive secretary for the International Society for MacIntyrean Enquiry. He is the author of *Reason, Tradition, and the Good: MacIntyre's Tradition Constituted Reason and Frankfurt School Critical Theory* (UNDP 2012), as well as numerous articles. Dr. Nicholas's research interests are varied, presenting and publishing on MacIntyre and critical theory, midwifery, love, and science fiction. He is currently writing a monograph on love as foundational for a political of liberation.

Christopher J. Orr is a PhD candidate in the Department of Natural Resource Sciences at McGill University. His work focuses on the political and ethical dimensions of transitioning to an ecological civilization. His doctoral research focuses on understanding deep transformations in society–nature relationships and explores their dynamics in the context of Canadian climate change politics.

Stephen Quilley is an Associate Professor in the School of Environment, Resources and Sustainability, University of Waterloo. Prior to this, he worked at the University of Manchester, the Moscow School of Economic and Social Science, University College Dublin and Keele University. Drawing on Norbert Elias, Karl Polanyi, and Ernest Gellner his research interests include: nation state formation and modernity; the long-term dynamics of human ecology; alternative political economy; ritual and ecological conscience formation. Dr Quilley is currently working on a SSHRC-funded project exploring a "pattern language for traditional music" as a driver of social capital and resilient community.

Piers H.G. Stephens is an Associate Professor in the Philosophy Department at the University of Georgia. He is the editor of the journal Ethics and the Environment, and philosophy reviews editor of the journal Environmental Values, with research interests in environmental philosophy and the history of ideas, especially ideas of freedom, nature, and the good in liberal and pragmatist traditions. He has co-edited three books, *Perspectives on the Environment 2* (Avebury, 1995), *Environmental Futures* (Macmillan, 1999) and *Contemporary Environmental Politics* (Routledge, 2006), and regularly contributes to journals including Environmental Ethics, Environmental Politics, Ethics and the Environment and Environmental Values.

Joshua Sterlin is a PhD student in the Department of Natural Resource Sciences (Renewable Resources) in the Leadership for the Ecozoic program at McGill University. He received his BA in Anthropology from McGill University in 2013. He went on to pursue an MSc in People and Environment (Anthropology) at the University of Aberdeen, Scotland.

Morgan Tait is a Philosopher of Science with an enduring interest in confirmation theory, whose work focuses on the interplay of the natural and social sciences and their social legitimation and epistemic authority in the domain of environmental governance. Tait holds a PhD in Philosophy, specializing in the philosophy of physics and philosophy of science from Western University. After completing his PhD, Tait accepted a post-doctoral position in the School of Environment and Resource Studies at the University of Waterloo, where he is currently a lecturer and adjunct professor in the Faculty of Environment.

Iván Darío Vargas Roncancio is a PhD candidate in Natural Resource Sciences at McGill University, a lawyer, and has Master's degrees in Bioscience and Law, and Latin American studies. Iván works on Amazonian legal cultures, the ontological and political dimensions of ecological conflicts, critical legal studies, and ecological law. His current research ethnographically follows indigenous practitioners, scientists, legal scholars, and ritual plants across territories, laboratories, and courts of justice in the Andean-Amazonian region of Colombia.

Rafael Ziegler is an Associate Professor at the Institut national de la recherche scientifique, Montreal, Canada, which he took up after ten years of directing GETIDOS. His work focuses on social innovation in relation to water, justice, and sustainability. After studies in philosophy and economics, Ziegler co-founded in 2009 the social-ecological research group GETIDOS (www.getidos.net) based in Greifswald and Berlin. The group jointly published *Social Entrepreneurship in the Water Sector— Getting Things Done Sustainably* (2014). He remains a GETIDOS fellow, and in pursuit of its important yet elusive maxim: GEtting ThIngs DOne Sustainably.

1

INTRODUCTION

Bruce Jennings, Kaitlin Kish, and Christopher J. Orr

The crossroads of progress

In geology and geophysics, the "Anthropocene" is one term that has been proposed to designate the fact that we now reside in a new epoch in Earth history, marked by large-scale and enduring changes brought about by human activity. During the Holocene epoch of the past 12,000 years, the climate of the earth has been unusually stable and moderate. This has facilitated abundant plant life and other environmental parameters within which the human species has vastly expanded its population, the range of its habitat, and its impact on ecosystems. The Holocene has provided a natural background for the biological evolution of human cooperation, and for the cultural evolution of complex forms of communication, social organization, adaptive intelligence, and behavioural plasticity.

Standing behind the impulse to declare a new epoch is the fact that human activity – which has always been biologically and ecologically symbiotic with other species – is now deleteriously transforming organic and biophysical systems on a planetary scale. Radical biodiversity loss and climate change are primary examples. These measurable, significant, and enduring anthropogenic effects are due primarily to the development of rich fossil carbon energy, new chemical and biological technologies, and complex forms of economic activity that have led to exponential growth in the human planetary footprint. Through feats of engineering, technology, and social organization, it has become possible for human beings to significantly control the material conditions of their lives. During the last 200 years at least, the arc of human history has been a movement from circumstances of natural necessity to those of increasing cultural agency and freedom – human social (and even individual) choice and control. This has been an uneven development, to be sure, but in the twenty-first century it has spread quite widely across nations and peoples around the world (Pinker 2018; Rosling 2018).

From this general conception of historical liberation for the human species, more specific ideals and rights of liberty or freedom for human individuals have been enacted in political, economic, social, and intimate forms of life. The ideologies and movements associated with liberalism, democracy, and entrepreneurial capitalism have contributed to this emphasis on individual liberty. Liberty is often defined as self-directed individual agency, free from arbitrary interference by others, in the pursuit of the subjective satisfaction of needs and desires. The right to liberty so defined is a moral claim that can be made by a person against others who are able to impede, impel, or coerce the person's behaviour in ways that conflict with the person's own purposes or interests. Private individuals may impede one's liberty, of course, but the right to be free quintessentially applies against those who wield corporate institutional power or the legal police power of the state. To put the point in a different way, liberty is territorial; it is the claim to be allowed the space for effective, self-directed individual actions in pursuit of the satisfaction of needs and desires; it is the right to live one's own life in one's own way, without outside interference or trespass.

The "others" who interfere and against whom the right to freedom of action is asserted need not be human beings, they can also be non-human beings or natural conditions that interfere with the individual's attempts to meet needs or fulfil desires. This point is important because claims have been made against non-human species and ecosystems on the grounds that their natural condition is impeding human liberty. Some have taken this rights-claim of liberty to ethically justify human actions to reengineer natural systems and to expiate unwanted species.

However, this trajectory of human progress that has inspired so much political and ethical aspiration in the modern period may have run its course. The recent realization that the modernity of Enlightenment liberalism, capitalism, and science and technology have created a world in which anthropogenic determinants and "forcings" can cause – and are causing – fundamental changes in planetary systems stands as a paradox of human historical development: the movement from necessity to freedom has reached biophysical limits and has become self-defeating.

Posing this paradox demands a reassessment of the concept and ethical importance of human freedom or liberty. There are two reasons for this, one motivational, the other structural. First, for at least 300 years, liberty has been one of the primary justifications for promoting the economic behaviour that has brought us to this point. Second, an entire social imaginary and powerful ideal has been built around the liberation of humankind from constraining necessity and unmet need. This ideal has given legitimacy to the economic institutions and legal rules that have facilitated and enabled such unsustainable economic behaviour and systems. Perhaps human beings have always dreamed of liberation and aspired to freedom from domination and want, but only in the modern period of Western

European and North Atlantic thought has this ideal of liberation been so strongly bonded to the political economics of mercantile, industrial, and now post-industrial capitalism, and to correlative notions of consumerism, psychologically endless material wants, and possessive individualism.

The twilight of liberal rationalism

This conception of liberty grew out of social and cultural practices developing unevenly among various class and status groups. Later, it was more explicitly and systematically given expression in philosophic and legal theory as "liberal rationalism" (Unger 1975; Spragens 1981). Liberal rationalism posits freedom as a defining essence of the individual human being, a natural right conceived universally and abstractly. However, liberal rationalism has largely theorized liberty in a way that renders invisible the fact that its own preferred conception of freedom was itself the product of particular social formations. As such, it is prone to be projected onto historical and societal settings in which it distorts local traditions and value perspectives, and often serves as an ideological cover for external intervention and political or economic control. At the same time, the interpretation of liberty provided by liberal rationalism has historically been associated with an atomistic, individualistic conception of the self and with a form of agency that involves the exercise of extractive power in the interactions between the self and others and between human beings and the natural world. Thus, liberal rationalism provides important ideas about liberty that must be tested against the political and ethical needs of the Anthropocene.

For this reason, in addition to the integration of freedom and obligation noted earlier, political theory needs to synthesize within the concept of liberty a universal abstract ideal and a practice rooted in relationships and concrete historical and cultural forms of life. It is not clear that the liberal rationalist theory of liberty successfully achieves this synthesis. Indeed, its idealized character makes its discourse vulnerable to co-optation by the highly abstract workings of the global financial and commercial systems. These systems are implicated in the ecological crisis and are seemingly insensible to any conditions or scientific analyses that are value-based. As systems, they employ purportedly value-neutral steering media (power and money) cut off from communicative deliberation and argumentation oriented at normative consensus, growing out of practices of solidarity and membership in civil society (Habermas 1987). By contrast, the steering media that direct Earth systems science and climate action politics are normatively explicit and are rooted in the tangible and concrete. Rapid and long-lasting changes in the biosphere, the atmosphere, and the oceans are not abstractions, but all-too-entangled situations for living creatures in ecologically struggling places.

If the material constraints on reasonable human agency finally do impinge on the cultural historical narrative of human liberation, political and ethical principles and ideals must be rethought in theory and enacted in new ways in practice.

The unrelenting and accelerating growth of anthropogenic interventions in biologic and geophysical systems has ceased to be the foundation for the flourishing of human life, and has become instead the principal threat to the continuing prosperity and well-being of our species. Scientists have identified nine essential planetary systems and proposed boundary limits on human activities so that we will not undermine those systems upon which all human societies and cultures depend. We may already have exceeded the safe operating margins of four of those nine systems (Steffen et al. 2015; Rockström and Klum 2015).

Our current social and economic systems drive this problematic human activity and do anthropogenic harm to Earth systems. This places the systemic ecological effects of human behaviour and human social and technological development in a new ontological light. To be sure, geologic and ecological upheavals punctuated the stable Holocene and brought massive changes to human societies within limited regions from time to time in the past. But never before have human beings caused such widespread disruption and destruction; never before has humankind understood so clearly that it is controllable anthropogenic activity that is causing these disruptions; never before has our species faced the imperative of undertaking such a deliberate, reasoned, and rapid transition in our economics, politics, social relations, and cultural beliefs and values.

For all its shortcomings, the liberal rationalist interpretation of liberty, and more broadly of human rights, has become a force in the domain of global ethics, deeply influencing much thinking about progressive approaches to law, property, work relations, and ways of life throughout the world (Inglehart 1990, 1997). Thus, for those who are heirs to this historically and sociologically determinate conception of freedom, it is difficult to achieve the moral integration and synthesis of freedom and the common good, let alone practice ways of life that resolve this tension.

And it is particularly difficult to win through to an ecologically bridled – and thus materially sustainable – understanding of freedom; indeed, ironically difficult at precisely this Anthropocene moment when we need such an understanding the most. Undoubtedly, in the Anthropocene, humans will face innumerable limits to their activity. But beyond the recognition of biophysical limits, the socio-political landscape ushered in by the Anthropocene will significantly change the way an individual functions day-to-day. This means more uncertain, less clear, and more chaotic understandings of liberty and agency. It is therefore difficult to imagine the kinds of changes we might see to rights, liberty, agency, and freedom as we move through the next few hundred years; they will challenge and shake the foundations of modern liberal society.

It is as if humanity stands poised before two buttons, one is an economic and cultural reset, the other triggers a self-destruct sequence. As a community of nations, we can't seem to agree on which is which. And even if we did, we don't seem to have the collective political will to stop those who have their finger on the self-destruct button and seem intent on pushing it, in order, they say, to protect our liberty.

Freedom and responsibility

The notion of the Anthropocene suggests that we have entered a period of transition in Earth history. This makes it imperative to revise common understandings of how humans should interact with natural systems and to acknowledge their inherent agency and ethical standing. Mainstream thinking in science, economics, and politics, particularly in the West, have regarded ecological and geophysical systems as a backdrop to human behaviour, a backdrop that does little but provide resources and raw materials for human use. Anthropocene awareness is a paradigm shift that involves seeing human beings in interdependence and symbiosis with non-human nature. It finds new dynamism and agency in the non-human world, and it broadens human ethical responsibility commensurate with the planetary scope of the consequences of human power and intervention.

The foundation of such a broader remit of human ethical responsibility is the proposition that nature has intrinsic value and is not merely a resource to be used for human interests. The conditions of its flourishing and resilience have a value of their own that human beings should recognize and respect; nature has interests and rights of its own that should be taken into consideration in the legal and moral governance of human agency.

This will place a moral restraint on human liberty and free agency in addition to the material limits and restraints that Earth system science is regularly discovering and reporting. That aspect of modernity that we associate with the intellectual movement of the Enlightenment has taken for granted that increasing knowledge leads to increasing human freedom and control. Contemporary scientific knowledge has called that assumption into serious question, and if we are to solve the paradox facing the future of humanity in the Anthropocene, then we must face the fact that many forms of activity and ways of life that have been pursued in the name of human freedom are dead ends, and that more ethically and ecologically viable ways of life must be constructed and practiced. In the long run, these new ways of life may come to be associated with the pursuit and the experience of freedom, but in the transitional short-term period that lies just ahead, they will be experienced as constraints that are incompatible with liberty as defined by liberal rationalism. If the ontological project of the liberation of human agency is to escape its current ecological impasse and contradiction, the historical project of human liberation must also be politically and ethically transformed.

Redefining human liberation

This ecological imperative complicates a fundamental issue in political theory: the integration and synthesis of the requirements of rights and obligations. In this regard, political theory and ethics walk a fine line between a libertarianism that is overly expansive and tends toward anarchism and the denial of legitimate authority root and branch on one side, and a conservative deontology or duty-based view in which liberty is unjustly constrained on the other. In any political order, particularly in a popular or democratic one, rights (such as the right to be a free and self-directing person) and duties (such as the obligation to respect others and promote the common good) must be reconciled; reason requires that ethical responsibility and the pursuit of liberty be understood as complementary, not antithetical. The basic idea is this. The fulfilment of ethical obligations cannot require moral slavery or the negation of liberty; if it does, then it is not a valid obligation that is being fulfilled but something else (e.g. power) that is being obeyed. Similarly, liberty cannot be fulfilled in the course of violating obligations to others; if it is, then it is not liberty but something else (e.g. narcissism) that is being served.

Yet matters have become more complex, because today this perennial problem of human ethics and politics is being reprised in the context of a newly perceived problem of planetary stability, and out of that, perhaps, the emergence of a new ethical worldview. Can the environmental conditions capable of supporting flourishing biodiversity and resilient ecological complexity be preserved now that collective human activities of production and consumption have reached intolerable and unsustainable levels? How then might the project of human liberation be reinterpreted, and along what lines might it be transformed? How can the hold of the liberal rationalist paradigm be broken?

Drawing on the work of the Canadian political theorist C. B. Macpherson and others, the following notions might point us in a productive direction. Macpherson's conceptual work on the kind of liberty that can claim the status of a moral right begins with the observation that unconstrained individual freedom of action that appropriates value from the materiality of nature, and from the labour power of human beings, is actually unfree at the structural level of society. And the self as the successful accumulator and expender of value and wealth is quite unfree at the structural level of personality. Such a critique opens the door to an alternative, democratic practice of freedom and ideal of the self. The conception of freedom presented by liberal rationalism is the idea of "negative liberty," or the absence of impediments to the fulfilment of individual will and desire. This notion of liberty is inadequate because it only rules out coercion, brute force, and other non-consensual constraints on the individual agent, without also taking into account the more subtle and insidious ideological or circumstantial pre-structuring of possibilities that individuals perceive and attempt to pursue. This foreclosing of choice is often done in the service of interests other than the individual's own, and should be seen as a morally significant curtailment of liberty in its own right (Macpherson 1973). The types of power that curtail liberty in this way are various, indirect, and in some societies pervasive. Macpherson refers to this as "extractive power," and it is actually facilitated by liberal rationalism and the concept of negative liberty, especially under the conditions of a market society and a capitalist political economy.

These considerations suggest that what an adequate concept of liberty requires is not non-interference in a narrow sense, but a form of life within institutional activities and power relationships that afford each person what Macpherson calls "immunity from the extractive power of others," or "counterextractive liberty." This idea opens the door to a related notion, namely, protecting natural systems and planetary boundaries against human extractive power. The notion of freedom as a condition in which domination and undue manipulation of wealth and labour are absent is a notion that could bring the agenda of social justice together with Anthropocene environmental justice and responsibility.

The notion of liberty and the conditions conducive to responsible agency and genuine liberation in the Anthropocene – call it "ecological liberty" or the liberty of ecological agents – will, we believe, have some of the following characteristics. Ecological liberty is freedom through relational transactions, ties, agreements, communities, caring, and sharing. It starts from empirical research, lived experience, and insight concerning the social being and the ecological interdependence and symbiosis of human persons (Gergen 2009). This social being is rooted in the human experience of caregiving and care-receiving that responds to human biological and psychosocial need, developmental potentiality, interdependency, and mutuality.

Human thought and behaviour are mediated through language and other systems of symbolic meaning, and this permits the creation of complex communities of reciprocity, cooperation, and normative order. Liberal rationalism acknowledges independence, but denies interdependence. Thus, it sees in this view of social being a threat of domination. Ecological notions of liberty reject the notion that an individual can be abstracted from social being and ecological place (Smith 2001). Staving off the preconditions for the possibility of liberty is not a defence of liberty; it is a negation or denial of it. The ecological interpretation of liberty provides a way of correcting the excessive atomism of many individualistic perspectives that have been and continue to be influential in liberal capitalist societies. However, despite its clear emphasis on the emplaced, situated, and communal aspects of forms of life and self-realization, ecological liberty does not point toward collectivism or even toward authoritarian and paternalistic versions of communitarianism.

Ecological theories of liberty tend to favour normative conditions like equal dignity and respect for individuals and democratic voice and participation for all. These theories see equality and democratic power-sharing as inherently bound up with the concept of liberty properly understood, not as competing values or principles that need to be prioritized or traded off against one another like "preferences" or "utilities" in the fashion of mainstream economic thought. The question is how to create a social order and forms of life in which liberty and other cognate norms, such as care, mutuality, and solidarity, can become mutually constitutive of one another. In this sense, a society of free, individual agency stands as a clear, critical negation to social orders and arrangements reliant on domination, exploitation, coercion, violence, seduction, or duplicity, each of which effectively reduces human and non-human beings alike from the conditions of subjects to the conditions of objects. Neither in relationship with other subjects nor when using natural objects is extractive power ethically justified, nor are human beings completely at liberty to ethically do what they will.

Ecological liberty cannot exist within the context of unjust structures of vastly unequal power, wealth, social opportunity, health, and psychological integrity any more than effective human economic activity can exist amid the degradation and breakdown of geophysical and bioenergetic systems. This provides a criterion for evaluating which types of transactions/interactions are to be nurtured, facilitated, and promoted by common rules and public policy, and which are to be discouraged or prohibited. The notion of parity of social membership and participation points to the communal or ecological dimension of relationship that informs an ecological account of liberty. From an Anthropocene perspective, non-human beings must be drawn into the orbit of moral categories heretofore reserved almost exclusively for human relationships. The challenge of reinterpreting and expanding the concept of membership by recognizing non-human beings as falling within its norms is to design human activities in such a way as to make voice and participation for non-human species and ecosystems meaningful in decision-making about human behaviour that affects the ecological common good.

Let us think carefully about how concepts drawn from moral and political philosophy can be interpreted to help meet the fundamental ecological challenges of the Anthropocene. When we continue to advocate for freedom in societies of the future, they will not be defending an atomistic, isolated value divorced from the interdependence of social justice, membership, and solidarity. And it is unlikely that human economic activities will ever be made compatible with the sustainable health of ecosystems and safe operating margins for planetary systems unless the notions of membership and solidarity are extended to the natural world. Liberty is the free agency of ecological selves extending recognition of membership to non-human beings and forging solidarity with nature.

Overview of the book

Our objective in bringing together the diverse perspectives in this collection is to build upon and critique existing views of liberty and ethically protected freedom of action, but also to bring these into conversation in ways that engage with the impending global ecological crisis and its implications.

The chapters in Part I provide a broad perspective by identifying some of the most serious and pressing problems that mainstream conceptions of liberty and the tradition of individualistic and self-interested agency will increasingly face in the ecological crises of the Anthropocene. In Chapter 2, Stephen Quilley leads off by arguing that conceptual binaries associated with the left–right political spectrum – within which virtually all definitions and debates about liberty have been imbricated – have evolved and can only be understood within a context in which economic growth and increasing complexity were taken for granted. Thus, a prosperous descent will require a high degree of political and cultural flexibility and will necessitate ongoing collective conversations to move beyond the traditional ideological camps of green, socialism, liberalism, and conservatism. Those conversations may help us navigate wicked dilemmas, see the adjacent possible, and prefigure alternatives that might bring about more ecological ways of being. And such conversations can create opportunities for greater happiness and well-being.

Steven Mock in Chapter 3 explores the historical relationship between nationalism and economic growth, arguing that understanding these relationships helps to explain instinctual conservative resistance to environmental proposals. He develops Ernest Gellner's ideas, which discuss how the nation evolved as the form of social organization best suited to managing the growth economy. Indeed, modern concepts such as liberty and agency are only comprehensible within nations which function to deliver economic growth. Mock's discussion provides an important version of the question: how can we preserve the values of modernity such as liberty, equality, and social mobility in a non-growing economy?

In Chapter 4, Kaitlin Kish applies complex systems philosophy from the field of socio-ecological resilience to understand systems transformation over time, arguing for the potential of prefigurative anarchist politics. She argues that capitalist modernity has created a highly resilient rigidity trap that upholds a powerful illusion of freedom while enforcing a mono-culture of beliefs and approaches to life, but prefigurative politics creates an opportunity for creative renewal within the system. Prefigurative politics allows for creative self-organization in an anarchic setting. As nineteenth-century anarchist Pyotr Kropotkin argued, one must "sow life around you" to be one with other, equal, and sharing. She finishes with an example of women as an opportunity for strategic self-organizing prefigurative activists, uniquely positioned to sow life.

Part II draws on diverse modern, pre-modern, and alternative traditions, which provide the inspiration for more ecologically grounded interpretations of human freedom and agency in both a social and an ecological context. These chapters develop notions such as relationality, non-domination, and the common good, pointing towards the need to re-negotiate human–nature relationships. In Chapter 5, Amy McCready examines the relationship between freedom and dependency, showing that dependency on nature is not necessarily at odds with freedom. She begins with the premise that knowledge and ethics are contextual, arguing that the dominant anti-ecological interpretation of freedom – that of self-sufficient rational actors – was forged in the patriarchal context of capitalist culture. Next, she shows that Locke's view of freedom implicates notions of embodiment and dependency, and draws on the relational ontology of feminist theorists to show how they entail relational obligations to both others and the environment. For McCready, relational dependency is a fundamental quality of life that brings with it obligations of responsibility. She proposes three interrelated modifications to the ethical foundations of liberal democracy: recognition of obligations to the relationships and communities that precede us; broadening our moral horizons to protect the vulnerable – be they social or ecological; and the need to root out systemic injustices and conceptual and ethical blind spots.

In Chapter 6 Peter Cannavò draws on the civic republican tradition of political thought to provide another strategy for reconciling ecological limits and liberty. He argues that they are compatible if liberty is understood structurally as collective self-rule. Cannavò connects liberty and limits using two key concepts: vulnerability and non-domination. Non-domination is central to the civic republican tradition, while in republicanism vulnerability and dependence on one another "is the fundamental starting point for any sort of politics." This extends to human dependence on nature. Republicanism attempts to cultivate human virtues and potentialities but recognizes the limits and vulnerabilities of this approach. He argues that attention to natural constraints and limits actually enhances freedom as collective self-rule.

Anna Beresford continues the discussion of the republican tradition and alternatives to liberal conceptions of freedom in Chapter 7, arguing that we must establish what flourishing means and what the good life looks like in our present age. She proposes that an Aristotelian eudaimonian virtue ethics engages with an ontological understanding of the human species that contrasts starkly with rational individualism and thus provides a way to develop an ecologically responsible concept of freedom. This rejects the ontology of the autonomous individual as a rational free agency, which rests on ontological assumptions such as individualism and the separation of facts and values. In contrast, virtue ethics adheres to a relational ontology and embedded morality that is compatible with an ecological worldview. She then proposes that virtue ethics entails a view of freedom understood as human flourishing in relation to and embedded in the ecological systems on which we depend. Freedom, then, is "a condition of the will arising from our nature being in the kind of world that we inhabit."

In Chapter 8 Jeffery Nicholas turns to an influential contemporary virtue theorist, Alasdair MacIntyre, extending his understanding of freedom as collective self-rule to show how freedom in that sense can be expanded from the exclusively human social realm to consider and include nature. He then examines these ideas in the context of protests by Native Americans and their allies against the Dakota Access Pipeline, showing that to move beyond assumptions of domination of nature, we need to expand the free community of collective self-rule to include nature as an agent. We are left with the challenge of collective self-rule, asking who stands for mother nature – Uŋčí Makhá?

Rounding out this section, Chapter 9 by Piers Stephens explores the relationship between nature experience and liberty. Stephens draws on pre-modern agrarian and sylvian traditions in which liberty is understood as a condition of human flourishing interacting with and respecting the rhythms of the natural world. Central to Stephens' argument is his ontology of nature, which regards naturalness as a spectrum, from something that remains relatively untransformed by objectifying, instrumental human rationality to relatively instrumentalized artifacticity. Stephens argues that we can continue to protect and enlarge our freedom through nature experience because it provides vital options for transformative growth and human imagination that expand human liberty as flourishing. Stephens' argument provides a compelling reason for why those who value liberty should not only protect but engage with and experience nature.

Moving from conceptual history and interpretation, the chapters in Part III examine practical aspects of liberty and agency in the context of modern economic and social life. In moving from an ecologically destructive to an ecologically benign civilization, they show how the groups, institutions, and practices that have claimed legitimacy on the basis of the liberty they afford in the past may give way to new institutions and different ways of living that provide social and personal freedom in correspondingly new ways. In Chapter 10, Rafael Ziegler leads off by locating freedom in a plurality of modes of provisioning beyond merely the freedom of consumption choices in a market. Consumptive freedom actually confines citizens in a vicious cycle of increasing consumption, decreased leisure time, and lack of meaningful choice or control. Instead, Ziegler argues that people must liberate themselves from affluence, reducing their consumption as the path towards genuine enjoyment. However, this individualistic view of change is sociologically naive without a structural transition towards a post-growth society. Growth-based capitalism undermines and eliminates our ability to engage in other modes of provisioning and thus structurally constrains our freedom to less meaningful choices. Plural modes of provisioning enable freedom while the dominance of market-based consumption eliminates freedom of meaningful choice. In a post-growth world, a plurality of modes of provisioning enables greater freedom from affluence but also freedom to make meaningful choices.

Offering an internal critique of neoclassical economics, Morgan Tait focuses in Chapter 11 on the limits of their assumption of rational agency in the Anthropocene. Tait argues that neoclassical economics maintains disproportionate influence and legitimacy in modern societies because it adopts foundational assumptions of physics for its epistemic authority. However, the epistemic authority of physics does not really apply in empirical economics, and it is only followed opportunistically. Neoclassical economists often retreat to humanistic arguments about the idealized nature of the rational agent when under attack. This duality is "cognitively unstable," and should undermine the intellectual authority of neoclassical economics in matters of governance and coordination. Mainstream economic theory becomes less and less relevant, even according to its own assumptions, as governance shifts to the scale of Earth systems.

In Chapter 12, Joshua Sterlin explores how we might see the global ecological crisis not as brought on by a universal humanity, but as the product of a certain type of dominating relationship characteristic of agricultural society. The crux, for Sterlin, is that some humans feel compelled to dominate and take liberties with the agency of other beings. At the root of the global ecological crisis is the transition to an agricultural civilization in which humans began dominating and forcing the Earth to give rather than receiving from a giving environment. Underlying this argument is the recognition that there is an intimate connection between ecocide and ethnocide, with every extinction of lifeways making humanity poorer, but also less resilient and able to relate to nature in different ways. In this way, anthropology has an important contribution to make in helping us see the myriad ways of being and can provide inspiration for diverse ways of understanding and relating to nature.

In Chapter 13, climate activist Aaron Karp argues that redefining freedom presents an important leverage point to transition from the dominant growth-based economy towards a steady-state economy. For Karp, core cultural concepts such as freedom are important sites of contestation between fundamentally different visions of the good life and ways of organizing society. He shows how, in the United States, corporations and mass media have constructed the dominant view of freedom as consumption, which is associated positively with free enterprise. He argues for more ecological and democratic definitions of freedom that recognize limits and foreground democracy. He then explores potential ways climate activists can help to redefine freedom. Karp concludes by proposing a range of strategies for how the climate movement might help to stimulate a societal discussion to redefine freedom and, in doing so, reframe climate change as an issue of freedom.

Part IV contains chapters that focus on the conditions of flourishing and sustainable life on the other side of the current crisis. In various ways, the authors in this section ask: what have we learned about being human and about what nature requires of us from our growing scientific and cultural awareness of ecological crisis? Are we coming any closer to a new horizon of harmony and reconciliation with the Earth? Policy studies enter the conversation in Chapter 14, as Martin Hensher uses the concept of optionality to explore and evaluate a range of alternative pathways which seek to guide humanity through the Anthropocene era. Optionality is the extent to which different pathways "maintain or maximize future freedom of action for humanity collectively, individually, and over time." While we cannot guarantee that deccelerationist pathways will avert collapse, he argues that accelerationist visions such as transhumanism and ecomodernism also convey no guarantees and may bring with them much greater risks of ecological collapse. Optionality analysis provides an important lens through which policymakers may evaluate alternative pathways to avoid ideological blind spots. And if the Anthropocene requires constrained pathways, Hensher suggests that optimizing choices on those pathways will represent an integration of the value of freedom and other values, and that it will be an ongoing exercise of agency and genuine freedom.

Stephen Quilley argues that concepts and subjective experiences differ dramatically depending on the level of social complexity and ecological imprint of their society. In Chapter 15, he explores the implications this insight has for liberty in the Anthropocene. He argues that "People really are different in different societies/historical times," corresponding with the level of social complexity and ecological signature of that society. Fundamental changes in complexity implied by a society with reduced ecological impacts would bring about radical changes in social relations and individual personality. Considering concepts such as liberty and agency in relation to the timescales of the longue durée implied by the Anthropocene framing, he presents a thought experiment, examining different conceptions of liberty and agency in different eras, and argues that this can help us envision plausible ways liberty and agency might change in the Anthropocene.

In Chapter 16, Iván Darío Vargas Roncancio explores how a new ecological jurisprudence might recognize and make space for the legal agency of non-human beings. He challenges traditional Western notions of legal personhood, which reinforce the separation between nature and culture. He argues that law needs to be adjusted to the logic of life, rather than trying to fit nature into our legal frameworks. He shows how relational ontologies of indigenous thought create space for non-human beings and their agency, for example, by foregrounding the co-emergence of humans and non-humans. Where indigenous cultures are given legal protection and voice, new questions in jurisprudence are opened up and will eventually spread to all areas of the law on a global scale: how do forests speak? How can we listen? When we do listen, how can the agency, rights, and flourishing of forests take their rightful place in parliaments and be represented before the law?

In Chapter 17, Christopher Orr and Peter Brown call for a heroic voyage of thought and action to bring into being a new human–Earth relationship grounded in a positive vision of the Ecozoic and tempered by educated hope. They argue that re-grounding the human–Earth relationship entails a foundational shift in the underlying ontology of modernity, and show how re-grounding the human–Earth relationship through such a seismic shift in underlying ontology has important implications for both agency and liberty. Using Thomas Berry's Ecozoic as the inspiration for a positive vision of a mutually enhancing human–Earth relationship – a state of harmony and resonance with nature – they show how liberty and agency play important roles in this vision. Being at home in the Ecozoic does not mean naive optimism for a return to the tranquillity of the Holocene; rather, educated hope can help guide humanity in responding with freedom and creativity to the spontaneity and natural rhythms of the Earth.

Conclusion

The modern period has seen a rich debate about the nature of liberty, its relation to the state and society, and the conditions attached to freedom. However, in the face of Earth system change today, these past debates appear incredibly narrow and all but oblivious to the sheer futility of pitting human liberty against natural necessity. We need to deeply reflect on our established ideas of both liberty and agency in light of this new context. Why? Because we don't just need to sustain a planet on which human life and non-human life survives; we also need to ensure that the natural and the human good (justice, freedom, the wild, and the flourishing of each species in accordance with its evolutionary capabilities) survives as well.

References

Gergen, K. J. (2009). *Relational Being: Beyond Self and Community*. New York: Oxford University Press.

Habermas, J. A. (1987). *The Theory of Communicative Action*, 2 vols. Boston: Beacon Press.

Inglehart, R. (1990). *Culture Shift in Advanced Industrial Society*. Princeton, NJ: Princeton University Press.

Inglehart, R. (1997). *Modernization and Postmodernization: Cultural, Economic, and Political Change in 43 Societies*. Princeton, NJ: Princeton University Press.

Macpherson, C. B. (1973). *Democratic Theory: Essays in Retrieval*. Oxford: Clarendon Press.

Pinker, S. (2018). *Enlightenment Now: The Case for Reason, Science, Humanism, and Progress*. New York: Penguin.

Rockström, J. & Klum, M. (2015). *Big World, Small Planet: Abundance within Planetary Boundaries*. New Haven: Yale University Press.

Rosling, H. (2018). *Factfulness: Ten Reasons We're Wrong about the World—and Why Things Are Better Than You Think*. New York: Flatiron Books.

Smith, M. (2001). *Ethics of Place: Radical Ecology, Postmodernity, and Social Theory*. Albany: State University of New York Press.

Spragens, T. (1981). *The Irony of Liberal Reason*. Chicago: University of Chicago Press.

Steffen. W. et al. (2015). Planetary boundaries: Guiding human development on a changing planet. *Science* 347, 6223 (15 February): 736; 1259855.

Unger, R. M. (1975). *Knowledge and Politics*. New York: The Free Press.

Navigating wicked dilemmas of liberty and agency in the Anthropocene

PART I
Navigating wicked
dilemmas of liberty
and agency in the
Anthropocene

2

LIBERTY IN THE NEAR ANTHROPOCENE

State, market, and livelihood

Stephen Quilley

> To be attached to the subdivision, to love the little platoon we belong to
> in society, is the first principle (the germ as it were) of public affections.
> It is the first link in the series by which we proceed towards a love to our
> country, and to mankind.
>
> (Edmund Burke, *Reflections on the Revolution In France*, 1790)

Introduction

The purpose of the chapter is to move away from categorical binaries of progressive and conservative thought in relation to problems such as gender equality, welfare safety nets, diversity, market efficiency – or liberty. I argue that concepts such as these, and their institutional operation at the level of policy emerged, and can only be understood, in a world of taken-for-granted economic growth. In this context, left and right understandings of the market and state were symmetrical and co-evolutionary. They depended upon each other at the level of discourse and policy instruments, but also in an oscillating pattern of adaptation to the evolving economy – a dynamic equilibrium which continually sought to balance incentives, investment competition, and innovation on the one hand, with social cohesion and political integration on the other. Moving out of that paradigm, into an uncertain and more chaotic world, requires that analysis steps back. Instead of offering firm prescriptions based on axiomatic moral and ontological frameworks of 'left' and 'right', a post-growth world is best approached by a consistent orientation towards 'wicked dilemmas' – i.e. paradoxical problems for which the trade-offs are complex and solutions are elusive. To the extent that habitual political orientations – feminist, socialist, liberal, conservative, green – remain in play and retain or regain traction, it may be because their meaning has changed fundamentally.

During the course of capitalist modernization, the domains of state and market developed in tandem (Figure 2.1a). The alignment of the left–right political spectrum has become naturalized so as to obscure these co-evolutionary dynamics: for many on the left, the extent to which the politics of social justice and emancipation depend upon an expanding market economy; and for many on the right, the extent to which the societal conditions for dynamic markets and technical innovation depend equally upon what Bourdieu called the 'left hand of the state' (Loyal and Quilley 2017). Capitalism and social democracy, in all forms, are handmaidens of modernity, both depending upon and promoting economic growth. All modern understandings of liberty (both negative and

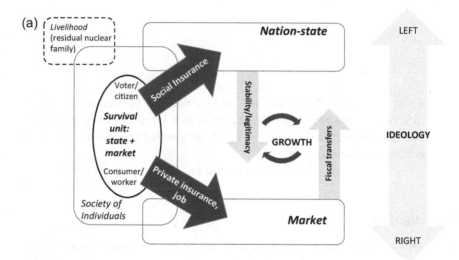

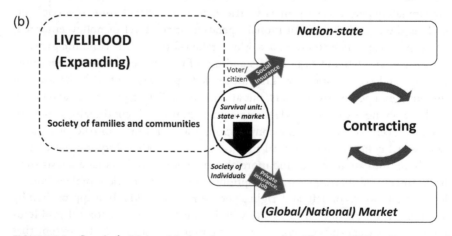

FIGURE 2.1   Survival Units: Market–State versus Livelihood: (a)The 'Society of Individuals' and the Market–State; (b)The Society of Families, Communities and Livelihood

positive – Berlin 2002) in practice hinge on the unprecedented degree of individual social and spatial mobility, and psychological individuation associated with complex, high-material/energy throughput, modern societies (Quilley this volume). Bearing this in mind, it is historically and ecologically significant that the growth of the state-market that made such societies possible was necessarily accompanied by the contraction of the domain of livelihood. In what follows, it is argued that ecological limits to growth are likely to see a reduction in the linked domains of *State* and *Market* and the partial re-emergence of *Livelihood*. Any such development may well be experienced negatively, as a reduction in the kind of liberty taken for granted by the spatially and socially mobile citizens of modern states. At the same time, in a post-consumer society there may also be opportunities for greater happiness and well-being consequent upon the re-integration of rights with social obligations, responsibilities and more re-embedded forms of economic exchange involving patterns of reciprocity (Figure 2.1b). Any such re-balancing of state and market with the domain of livelihood would transform the meaning of, and relation between, taken-for-granted ideological orientations such as feminism, nationalism, and conservatism, foreclosing some possibilities whilst opening others. Notably, with regard to the problem of liberty, it is argued that such a context would be conducive to: forms of associative socialism attentive to self-organizing communities (Hirst 2001) and sensitive to national-civic sentiment (Orwell 1962); feminist ideologies attuned to the domestic sphere as an undivided arena for both production and care (Kish 2018); and forms of conservatism critical of market liberalism, foregrounding the role of virtue and attentive to the web of connectivity fostered by Edmund Burke's 'little platoons' in civil society (Scruton 2015, Macintyre 2016). These potential overlaps are obscured by the increasingly anachronistic left–right lens through which the ideological landscape is still viewed. By stepping back and exploring the wicked dilemmas associated with the prospect of a lower energy, low-throughput modernity, it may be possible to navigate a freer and relatively more benign path through the Anthropocene.

Energy, complexity, society, psychology

It is widely recognised that social complexity is a function of energy throughput (Quilley, this volume; 2011, 2013; Tainter 1988; Odum 2007; Morgan 2018a, b). Echoing Marx, Elias (1978) showed that, in all periods of human development – language and fire, stone tools, agriculture, bronze/iron tools, industrialization, the information age – three sets of controls that co-evolved and developed in tandem, namely, controls over nature (*eco-genesis*[1]), controls over people (*socio-genesis*), and controls over self (*psycho-genesis*). In consequence, material changes in the pattern of life are associated with corollary changes in the mode of consciousness and thinking (Ong 1991; Barfield 1965). For example, peasant living generates characteristically peasant ways of thinking (about land, gender, nature, rights, obligations, deities, life, death) – a way of being that Marx rather rudely likened to potatoes in a sack, on account of the privatistic, familial mentality. Large factories were, he observed, more likely to generate such a collectivist class consciousness. It seems clear that, in the information age, the internet and social media are creating new material conditions which are fostering, in turn, very different patterns of consciousness; changing the neurological structure of the human brain, and the emergent networks through which minds are patterned; and ultimately transforming the experience of the self in relation to others. Such changes are a function of the ontogenetic growth and development of individual brains in very specific social and technological contexts.

From this perspective, we see that particular social–ecological regimes condition very specific societies, which engender equally specific structures of personality. It follows that the modern society of individuals (Elias 2010) must be understood as a direct corollary of the low entropy social complexity made possible by flows of energy associated with fossil fuels. Demographic growth, urbanization, and the extension of the division of labour hinged directly the capacity of agriculture to support a non-producing population of social and economic specialists and, most recently, the paradigmatic increase in the social and spatial mobility of ordinary people – a process that greatly enhanced the experiential and neuro-psychological drivers of individualization.

Nation-state formation

This 'society of individuals', that is a functional premise for any liberal–democratic or capitalist society, is not natural. For millions of years, in one form or another, tribalism has formed the basic architecture of human association (Weiner 2013; Fox 2011). Human beings are social animals and we cannot exist in isolation. Until very recently, social groups based on family/community/place constituted the most basic and exclusive 'survival unit' and, for the most part, ejection from such a group meant death. At the same time, human beings have an innate propensity for insider/outsider dynamics – a competitive instinct whereby projection of aggressive hostility and competition against a 'they' fosters in-group co-operation and loyalty.

With processes of social development and access to greater flows of energy through agriculture, the basic units of tribal societies increased in size, with family bands giving way to chiefdoms, more complex tribal associations and, eventually, agrarian states and empires. The nation-state however represented a step-change in political organization because, for the first time, it created a direct link between a 'we-identity' and institutional architecture at the level of the state and the individual citizen – a relationship, in principle, unmediated by intervening familial, tribal, or group identities. The emergence of the modern state was bloody and violent, always involving, in one form or another, the subordination of rival warlords to a central authority, the dispossession of peasants in the context of processes of urbanization and industrialization, and the emergence of a national economy dominated by 'disembedded' price-setting markets. A condition of existence for such states is the emergence of a unified 'high culture' based around (ideal-typically) a single dominant language. Since most nation-states emerge in the context of a myriad of potential and competing language-cultures that were extinguished as a matter of policy and state-craft, the process of state formation invariably involves an aspect of what is technically 'cultural genocide'. With this in mind, it is important to recognize that taken-for-granted liberties tied first to citizenship, and more recently to the concept of universal human rights, became conceivable and actionable through the lens of early-modern and Enlightenment philosophies, which were themselves only thinkable against the backdrop of emerging national societies and complex market economies. Paradoxical though it may sound, modern liberty is dependent on historical episodes (and, in some cases, ongoing patterns) of cultural genocide. For example, it is difficult to imagine the positive liberties associated with modern healthcare and education being achieved in Papua New Guinea, without the tragic and 'inevitable' loss of many – possibly the great majority – of that country's 800 language-cultures. Why should this be inevitable? Because modern market and state infrastructures require civic modernization and the emergence of a society of spatially and socially mobile individuals. This emergence involves the replacement of informal, inter-generational, family/community-based forms of socialization and education, based on experiential immersion and what Ingold refers to as the 'education of attention' on the one hand, with what Gellner called 'exo-education' on the other: i.e. formal processes of educating involving the inculcation of language, literacy, numeracy, technical categories, legal and behavioural norms, standardized categories (e.g. weights and measures, time) and transferable skills – all of which are prerequisites for the processes of market-led economic growth, modernization, and rationalization (Elias 2012b; Gellner 1983; Polanyi 1944; Ingold 2000; see Loyal and Quilley 2017).

Significantly, in this society of individuals, group identity does not disappear. Rather, the 'we-identity' is articulated at the scale of the nation. This national identity is of crucial importance, not least because it legitimates both the state's monopoly of violence and the institutions of the economy (currency, contract law, import tariffs, etc.), but also more generally the process of redistribution via fiscal transfers. To the extent that they fund an array of national infrastructures and entitlements, citizens become accustomed to seeing as legitimate tax appropriations that, in previous centuries and other contexts, were viewed as little better than armed extortion. The entire architecture of the social compact and public infrastructure more generally (e.g. health, transport, education, military, police, roads) in both Western states and emerging middle-income countries depends on the legitimacy of such transfers. It is for this reason that a central ideational mechanism in nation-state formation involves the projection of citizenship as a quasi-familial relationship. This 'imagined community' (Anderson 2006; Hobsbawm and Ranger 1983; Thompson 1991) is constructed, often very deliberately, on the back of a national mythology involving history, a constructed ethnicity, and the valorization of a symbolically shared culture (including food, music, poetry, ostensibly shared national traits, etc.). This we-identity is fostered and promoted at every opportunity, not least through the 'pretend war' that is involved in games such as football, rugby, and cricket. So, while in the course of a de-traditionalization (Beck 1992; Giddens 1991; Elias 2010) the I/We balance shifts decisively towards the 'I', this background we-identity of the nation provides a continuing condition for the social and spatial mobility of transacting, choosing, and self-determining individuals. For peasants in agrarian societies, group-identity was pervasive and shaped the perceptions, quotidian rhythms, and the fine-grained, weave and weft of the life-world. For modern citizens, the 'we' retreats into the background of society and the psyche. And yet it is there, always, legitimating the extant social compact and underpinning a set of internalized social commitments (to obey the law, pay tax, abstain from violence) that constitute the survival unit for individuals who have little recourse to extended family, community, or tribe in times of insecurity or hardship. Clearly this national we-imaginary is overlaid by other – ethnic, regional, class, religious – identities. And yet it is undeniable that the success of the nation-state has depended on the resilience of a national 'we' that proved as obstinately durable to the class warriors of the Second International in 1914 as it does to European technocrats in the era of populism and Brexit (Scruton 2015; Orwell 1962).

Modern liberties, individualism, and the nation-state

Here, then, is a paradox. Modern individuals and the institution of the market both depend on the modern state for their existence. Individualization – the spatial and social mobility of individuals, detached from traditional structures of mutual obligation associated with family, place, and land – may depend on the burgeoning market economy. Survival comes to be a function of the capacity of skilled individuals to navigate the transactional world of the labour market (as well as combinations of individuals to press for collective advantage and security). But the content and rules governing such transactions – contract law, currency, prohibitions on violence, a standardized economic and cultural milieu that facilitates individual movement – depend on the state. As Polanyi wrote, the state 'institutes' the economy (1957).

Furthermore, as Polanyi (1944) described, left to its own devices market rationality and the minimal 'night watchman' state (Nozick 1974) would destroy both society and nature. Faced by acute problems of social order that threatened the existence of the state (and haunted by the spectres of 1789, 1848, 1871, and 1917) and the more chronic erosion of environmental public goods (e.g. urban pollution), the utopian project of the free market was saved in the end by a 'countervailing movement for societal protection' – a process which started with the legalization of trade unions and some limited factory acts, state-sponsored education and social insurance schemes, and culminated eventually in the emergence of comprehensive Keynesian welfare states (Quilley 2012).

In this process, the I/We balance has shifted dramatically to the pole of the 'I'. Liberal societies are now associated with a very real and unprecedented freedom for individuals to move about, adopt occupations and social roles, and experience forms of individuation that are independent of ascribed social status, gender, sexuality, location, etc. This observation is significant precisely because it underlines the extent to which market and state developed in tandem, and at the expense of livelihood.[2]

Thus, the society of individuals notwithstanding, the innate propensity for established/outsider binaries and in-group identification (Elias 2008) or tribalism (Fox 2011) has not disappeared. The de-traditionalization and individual social-spatial mobility associated with liberal society described at length by sociologists such as Giddens (1991), has depended, thus far, rather completely on the 'we-identity' and in-group associated with the imagined community of the state. As a 'survival unit', the state is abstract and operates through a raft of intervening institutions. Except in times of national conflict and crisis, individuals do not access material support with reference to the mode of mutual identification associated with shared citizenship. But nevertheless it is that mode of mutual identification that makes possible the more mundane institutions of state and market, which provide the lattice of physical security and material support that allow liberal-democratic states to cohere and grow.

End of growth: Politics in the context of declining prosperity

This brings us to Tim Morgan's insight (2018b), that virtually all Western institutions are products of growth. He was not the first to arrive at this conclusion (see Tainter 1988; Odum 2007; Quilley 2011, 2013; Homer-Dixon 2007). Nevertheless, his formulation is pithy, and – since it comes from someone who has spent a career as a strategic analyst at one of the largest banks (Tullet Prebon) at the apex of the historical and current process of capitalist modernization, the City of London – it is perhaps significant in marking a moment when the energetic fragility of modern civilization began to percolate into elite consciousness.

I will comment briefly on the relation between the material consequences of declining growth and the possibility of non-linear disruption of the current landscape of political values and ontological assumptions. I explore these relationships through three ideological cornerstones of modern life: conservatism, socialism/social democracy, and nationalism. The objective is not to develop any fully fledged critique or endorsement of any particular ideological position, but rather to suggest that ideological frames co-evolve across an ideational landscape, and that any seismic rupture to this underlying landscape will present existential problems to certain 'ways of thinking' – but at the same time engender opportunities for hitherto unthinkable combinations of ideas, moral intuitions, and political coalitions. These potentials can be thought of as existing in an 'adjacent possible' (Kauffman 1995) area of the political landscape that is generally rendered invisible by the dominant architecture of institutions and ideas. It is from this adjacent possible, if anywhere, that the building blocks of an alternative, ecological modernity or a paradigmatically different mode of political economy, will be found.

Conservatism

Conservatism in the Anglophone world always contained two threads (Scruton 1984). On the one hand, there was social conservatism, characterized by the Burkean concern with tradition, community, and social cohesion, and rooted in a Romantic image of an organic 'gemeinschaftlich' rural society. This was predicated on the acceptance of hierarchy, but at the same time insisted on a lattice of mutual obligation from which no one, regardless of rank, was exempt. On the other hand, market liberalism was the forward-looking and progressive vision associated with Adam Smith, that centred on the efficacy of markets – a society of individuals in which the hidden hand, by some alchemical magic, transformed private vice into public virtue. This mechanism allowed market evangelists to discount traditional virtues and social restraints, confident that the aggregate individual rationality would both drive technical and social progress and solve problems along the way (see Scruton 2015; Beresford, this volume). Whereas social conservatism harked back to the virtue ethics and natural law of Thomas Aquinas and Aristotle, market liberalism was firmly in the tradition of the Enlightenment and animated by more utilitarian, rationalist, 'meliorist', and individualist commitments. The latter crossed over with the emerging tradition of liberalism and eventually consolidated a consensus that extended to include centrist social democrats. This consensus centred on a commitment to the 'society of individuals' as a necessary prerequisite for democracy, rule of law, citizens' rights and, later, human rights. American libertarianism is a radical and sociologically naive extension of this view, insisting that a society of individuals can exist without the state.

Modern conservativism, operates for the most part as a variant of classic liberalism predicated on a ubiquitous moral individualism. Aside from a general commitment to tried and tested institutions, such conservatism has all but left the radical vision of Edmund Burke behind. Certainly, any communitarian vision linking individual rights to mutual obligation and duty, and a Thomist/ Aristotelian understanding of natural law, has largely disappeared from the conservative lexicon in the UK and Canada. In the United States, such commitments have been sustained in the form of a fundamentalist Christianity. The landscape of Republican politics, and even the outlook of individual politicians is persistently fractured by the bubbling tension between secular market individualism and the utility of private vice on the one hand, and a morally prescriptive Christian ethics on the other (Kolozi 2017; Tait 2019). These wings of the party are, however, united by a consistent link between rational individualism, the sanctity of private property, and hostility to collectivism in the form of the state. Any non-state communitarian impulses are lost in this constantly looping drama of the individual David and the Federal leviathan Goliath. Conservatism in its neo-conservative, neo-liberal, and liberal variants represents Adam Smith on steroids and the almost complete eclipse of Edmund Burke.

However, if modern conservatism, like social democracy, is a product of the age of growth, how might it respond to an era of collapsing growth rates and declining prosperity? One very likely scenario is the resurgence of that older Burkean impulse towards tradition, family, and continuity, and the necessity of a lattice of mutual obligation, if not naturalized hierarchy. And, to some extent, the elements necessary for such a conservative communitarian revival are already in place – in the resurgent philosophical currents associated with virtue ethics and Thomist theology (Tait 2019; Millbank and Pabst 2016; Macintyre 2016); in radical Catholic discussion of the 'Benedict Option' – a prospective communitarian retreat from consumer capitalism (Dreher 2017); and in long-standing traditions such as Distributism, inspired by the writings of GK Chesteron and Hillaire Belloc (Lanz 2008).

One feature of the burgeoning populist insurrection that has shaken Western democracies since 2016 has been the significance of the fissure between a 'placeless' globalist-liberal-internationalist orientation on the one hand, and a nationalist sensibility reasserting the sacred community of particular places on the other. This tension runs through parties of both the left and the right – with the ironic result that in terms of the agenda around, for instance, open borders and migration, the Koch brothers (long time bête noire for the left) are closer to Alexandra Ocasio Cortez and the far left of the Democratic party than to Steve Bannon. This has not gone unnoticed (not least by Fox News pundit Tucker Carlson 2019). As the tension between these orientations grows, a space for far-right nativist movements has clearly opened up, particularly in the European Union. But conservatives are simultaneously rediscovering older commitments to civil society as an arena for a tissue of association, virtue, reciprocity, and familial and neighbourly obligation as the roots of a meaningful life (Scruton 2015). It is at least conceivable that, in an era of limits, this rediscovery of Burke's 'small platoons', along with a recognition of the excesses of corporate power and global markets, may open the way for remarkable and hitherto unthinkable conversations between conservatives, socialists, feminists, and greens about the appropriate relation between state-market and the domain of livelihood.

Socialism and social democracy

All attempts to realize the socialist vision – whether the radical commitment to the socialization of the means of production, or the social democratic commitment to the mixed economy – have started from a number of assumptions. These include: a largely taken-for-granted trajectory of technical innovation; the directive role of the state in regulating markets (at the least) and organizing (at most, the totality of) production; a primary or exclusive role for the state in (re)distributing wealth; a trajectory of unconstrained economic growth; a vision of human nature as plastic and therefore allowing the (gradual or immediate) re-formation of personality and behaviour in accordance with abstract norms of equality and (initially) socialist citizenship and, more recently, in accordance with an internationalist vision of human rights.

With regard to the problem of growth, it was only the miracle of the fossil-fuelled industrial revolution that made the idea of a complex society without slavery or servitude even thinkable. Growth was the default assumption of all Marxist theory and practice (Kolakowski 1978). And, as demonstrated by iterations of 'regulation theory', growth was also a prerequisite for all post-war developments in Western social democracy (Boyer and Saillard 2002). Certainly, from the 1970s, with growing awareness of global ecological crisis, eco-socialists in the tradition of William Morris's *News from Nowhere* and Callenbach's *Ecotopia* (1990), sought to advance a socialist route to sustainability (Gorz 1985, 2010; Lipietz 1995). Others argued that all the answers were already to be found in Marx (Pepper 1993). But these theories, and the movements they spawned, were stillborn, largely because proponents failed to address two central problems in socialist theory. Firstly, the state has proved an abject failure as a replacement for the market in the coordination of production and consumption and allocation of resources. Socialist attempts to address Hayek's (1945) 'information problem' have proved marginal and unconvincing (e.g. Wainwright 1994). Secondly, in common with sustainability advocates more generally, eco-socialist theorists have failed to show how high levels of social complexity can be reconciled with ecological integrity.

On the other hand, there are influential traditions on the left associated with the English socialism of William Morris (in his medievalist incarnation) and the anarchism of William Godwin (1793) that focus less on the state and more on the self-organizing capacity of communities. Their critiques are levelled at the disorganizing intrusion of both the leviathan state and the unconstrained market into the life world of organic communities. In the age of the Internet, this tradition is re-emerging with visions of highly distributed, decentralized and community-based forms of market production. Kevin Carson' s 'home-brew industrial revolution' presents the most complete articulation of these ideas (Carson 2010; Kish et al. 2016).

Whatever its merits in a world of growth, any vision of state-directed, redistributive socialism is untenable in a world of declining growth rates and economic contraction – *at least as a self-contained solution set*. To the extent that material inequality needs containing (for social cohesion), and the scale of human activities curtailing (for ecological reasons), a viable model of political economy will more likely involve forms of self/social constraint arising from a more viscous relation between rights and social obligations in the context of place-bound community and family networks (see below). Some aspects of the social democratic safety net may remain in place at the level of the state, but only if these find successful accommodation with lower-level forms of provision and infrastructure delivered through other means (in accordance with something like G.K. Chesterton's understanding of subsidiarity – see Médaille 2010; Millbank and Pabst 2016).

Nationalism and the imagined community of the 'nation'

As with modern conservatism, liberalism, and socialism, nationalism is a product of the age of growth, along with the architecture of the nation state (see Mock, this volume). If the energy and material flows marshalled by society ever decline such as to preclude an effectively functioning territorial authority with a monopoly of violence and the capacity to institute a national economy, then it would seem inevitable that the nation-state would give way to more local, and probably warring, tribal survival units (see Kish on 'localized anarchy', this volume). As Mock warns, such an eventuality would threaten a broad sweep of cherished civilizational goods, including democracy, literacy, mass education, human rights, and social mobility. And any such development would in time see the disappearance of highly individualised modes of consciousness and self-actualization, as well as a dramatic shift in the I/We balance towards the latter. Peasant living would likely engender peasant modes of consciousness.

Cataclysmic civilizational collapse in the wake of climate change or global war is certainly possible, but such a scenario need not detain us here. The important thing is to recognise the nation-state and the current impetus to supranational forms of governance, globalization, and cultural globalism are points in a continuum of societal organization and complexity (*sociogenesis*), co-evolving with material/energy flows (*ecogenesis*) and patterns of personality/self-formation (*psychogenesis*), i.e. the triad of controls (Elias 1978).

Declining growth rates are very likely to engender some reversal of the 1,000-year-long process of nearly continual complexification that is at the heart of the 'great acceleration'. At the very least, this suggests an imminent countervailing movement against corporate globalization and cultural globalism – the pronounced pattern of interdependent economy and institutions most evident in the project of the European Union. This process may be drawn out and take the form of systemic crises in the global economy and/or apparently regressive, nationalist political projects. The latter are already beginning to emerge with the series of populist insurrections that have dogged Western politics since 2016.

This process may have 'left' and 'right' wing incarnations and might be associated with more or less benign consequences. Any parties wishing to be influential in structuring or channeling the process will be forced, along the way, to reconsider many political assumptions and cherished values. This is because in the age of declining growth, existing institutional clusters and ideological commitments may no longer 'fit' together.

The idea of the nation is a case in point. For 200 years, the nation-state was the quintessential vessel for Enlightenment thinking and social democratic emancipation – allowing the loosening of the ascriptive ties binding individuals into traditional gender and class roles, whilst insulating such newly emancipated citizens from the vagaries of the market. The nation-state was progressive because it established citizenship as the basis of social solidarity and an 'imagined community' (Anderson 2006) that potentially allowed all members (regardless of colour, creed, etc.) a seat at the table. But over the last 20 years, the exclusive forms of solidarity associated with citizenship have become, at least in the eyes of many far-left progressives, politically 'suspect' and incompatible with the more universalist, internationalist, and utopian aspirations associated with the idea of human rights. It is not surprising, therefore, that the 2015 refugee crisis engendered a tipping point. Chancellor Merkel's decision to open the borders was embraced enthusiastically by parties of the left. But in the wake of that decision, millions of working-class voters in countries across Europe have switched their allegiance to populist parties. Although ostensibly 'right-wing', nativist parties, such as the Sweden Democrats, the Dutch Freedom Party, Marine Le Pen's (then) National Front, and the Alternative für Deutschland (AfD) have been much more vocal about defending the welfare state and the post-war social compact from both the free movement of capital (neoliberalism) and the free movement of labour (refugees and economic migrants) (Goodwin 2018).

As it turns out, the stabilization of a global architecture through which to make good the cosmopolitan aspirations of the left, requires too great an ecological price. Barring shifts in the human technological capacity to process energy and materials, the overall 'transformity cost' (i.e. the energy embedded in all the antecedent processes distributed across the entire network of activities required to produce any particular good or service – Odum 2007) of this level of complexity has proved too great. As a result, the end of the age of growth will see some degree of reversal beginning with the re-nationalization of economy and public life.

This is already beginning to happen, as open and connected societies struggle to contain the social tensions produced by over-exposure to global markets. Now supported by Italian populist leader Salvini, Steve Bannon's 'Movement' is one ideological expression of this retrenchment (Borrelli 2019). In order for such a process of re-nationalization to retain progressive and emancipatory institutions and ideas carried over from the age of growth, there *would* have to be a renewed emphasis on the active construction and consolidation of an 'imagined community', with a shared national story, including an origin myth, to coalesce a strong and visceral pattern of mutual identification among citizens. As indicated above, it is such a quasi-familial we-identity that makes possible a fiscal-welfare system based on legitimate patterns of redistribution. On the other hand, as Scruton argues (in line with social Catholic principles of subsidiarity), the difference between fascist and conservative approbation of the nation state relates precisely to the conception of sovereignty. Specifically, the difference turns on whether power and patterns of mutual identification are construed as emanating down from the state onto a subservient civil society, or upwards from the obligations and attachments of family and neighbourliness, through Burke's little platoons of 'purposeless associations.'

In the 'near bad future' evoked by the prospect of Anthropocenic collapse, it is unlikely that the state alone will suffice as a security and survival unit. There are very positive reasons for considering an entirely novel configuration in which contracting but nonetheless dynamic domains of state and market are rearticulated with a re-emerging domain of livelihood (see Figure 2.1b). Nevertheless, for obvious reasons, it would be preferable to sustain, as far as possible, a viable state capable of delivering infrastructure, some form of safety net (possibly a minimal basic income), legal system, and an effective and transparent (formal) market. For this reason alone, any kind of progressive politics in the future will depend on left and liberal constituencies rediscovering an affiliation with the idea of the nation state.

Liberty in the Anthropocene: The politics of livelihood

Unlikely as it seems, in this prospective world turned upside down, there may be incongruous overlaps between otherwise divergent and antagonist traditions – hidden potentials that might coalesce to form a very different configuration of economy and society (Figure 2.2). This latent potential is a part of the 'adjacent possible' that is difficult to express or formulate in a context of growth, but may come to be seen, in coming decades, as commonsensical and obvious.

What does the changing I/We balance associated with any post-growth society mean for liberty? As Morgan points out, familiar modern ideologies were born of the age of growth and rising prosperity. They are predicated on assumptions about continuing marketization, technological growth and individualization, and an inexorable rise in social complexity. From a thermodynamic perspective (Odum

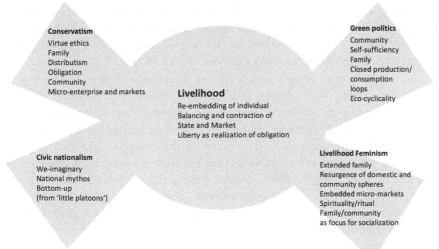

FIGURE 2.2 Unlikely Ideological Affinities of Livelihood

2007; Quilley 2011, 2013; Ophuls 1977, 2011), such assumptions aren't warranted. Any systemic shift away from economic growth and rising prosperity will initiate a progressive loss of complexity (Kish & Quilley 2017). The important questions will be how far this goes and how fast. In all of the ideological camps surveyed above, the background assumption is that cherished features of the Enlightenment – the sanctity of individuals, liberty, democracy, human rights, citizenship, scientific rationality – are non-negotiable (Quilley 2014, 2017). But, of course, as the saying goes, nature bats last. Any feature of society comes with a set of preconditions and an ecological and thermodynamic price tag, in terms of the associated throughput of energy and materials and the generation of waste and pollution ('disorder'). Nothing is sacrosanct, and everything is relative. Thus, forms of national identification that might seem regressive and anachronistic when compared with the majestic peaks of a global consciousness, may turn out, on the 'down slope', to be wonderful frameworks for stability and cohesion. In such a context, it would be a mistake to allow the idea of the nation state and nationalism to acquire a permanent and indelibly negative patina – the disavowal of the 'good' in the name of an implausible utopian 'best' (Hazony 2018).

The common thread, that makes this configuration conceivable, relates to both the value ascribed to the individual and the problematic or unsustainable nature of a 'society of individuals' (Elias 2010). Conservatism and green politics are both deeply committed to a vision of the free individual agency that coalesced during the Enlightenment, but which actually goes back to the Judeo-Christian tradition and was given impetus in both Protestant and Catholic traditions by the upheaval of the Reformation. Capitalist modernity has emancipated people, but, at the same time, has left individuals untethered and in a condition of anomie (Durkheim), disenchanted and caged by the implacable forces of rationalization (Weber), alienated and deprived of their 'species being' (the young Marx), and narcissistic, neurotic, unhappy, and self-absorbed (Lasch, Fromm, Freud, Jung). For green critics of modernity, individualization is linked to compulsive consumerism and a lack of restraint. For social conservatives, excessive individualism has led to degeneracy and the undermining of civil society.

On this basis, it is possible to imagine an alternative form of modernity in which the dyadic poles of *Market* and *State* are leavened and balanced by a re-emerging domain of *Livelihood* (Figure 2.1) (e.g. see Zywert & Quilley 2018 in relation to health systems). Although re-embedding individuals in local, familial, and community forms of association that provide significant degrees of security and safety, this scenario doesn't necessarily imply the disappearance of either the state or rationalized, price-setting markets – just that these spheres contract relative to the re-emerging domain of livelihood. Nor does it require that citizens are not able to access security on an individual basis through social and private insurance and provision. It does suggest that these forms of provision will no longer support a cradle-to-cradle security system, and that there will be an onus on families and communities to look after each other on the basis of informal or formal reciprocation. The partial move towards informal economy and reciprocity will be dictated partly by cost. In a context of declining growth, the fiscal capacities of the state will be permanently over-stretched.

Although there isn't room here to deal with the issue adequately, the aspect of the modern consensus that may be most challenged by the politics of (de)growth is feminism in all of its liberal and radical forms (see chapters by Kish, Mock, and McCree in this volume). Although the threats are hard to ignore (particularly in relation to progressive family law and welfare entitlements), there are also opportunities and perhaps unimagined configurations of work, family, and security – opportunities that Kish (2018) refers to 'polioikos feminism'. These are alluded to in Figure 2.2 as 'livelihood feminism'.

Finally, ecological limits will turn upside down the certainties of the left–right ideological spectrum. The re-emergence of livelihood and the re-nationalization and re-localization of the market economy will make possible forms of political economy and ideological framing that, currently, may seem quite inconceivable. There is no guarantee that these trajectories will be benign. It seems inevitable that, at least with regard to the expectations of social and spatial mobility that we now take for granted, such changes will be experienced by many as a reduction in liberty. Nevertheless, there are certainly more and less benign outcomes. A prosperous and peaceful descent will require a high level of political and cultural flexibility. Specifically, it will require a willingness for an ongoing conversation between liberals and social democrats and left-greens on the one hand, and social conservatives, civic-nationalists, and people of faith, on the other.

Notes

1 Although Elias did not actually use the term 'ecogenesis'.
2 Scruton (2015) makes the same point as Polanyi from a conservative perspective, cautioning against the impossible utopianism of Nozick's 'night watchman state (1974)' But whilst recognizing the need for a safety net, his instinct is for the revitalization of livelihood and an associative democracy in which individual mobility is at least partially constrained by personal and face-to-face associations with those 'small platoons' celebrated by Edmund Burke.

References

Anderson, B. (2006). *Imagined Communities: Reflections on the Origin and Spread of Nationalism* (Revised ed.). London; Brooklyn, NY: Verso.

Barfield, O. (1965). *Saving the Appearances; a Study in Idolatry*. New York: Harcourt, Brace Jovanovich.

Beck, U. (1992). *Risk Society: Towards a New Modernity*. SAGE.

Berlin, I. (2002 [1969]). "Two concepts of liberty." in I. Berlin (ed.), *Four Essays on Liberty*, London: Oxford University Press. New ed. pp. 118–172.

Borrelli, S. (2019). "Steve Bannon: Italian experiment 'will change global politics'." *Politico* Retrieved July 4, 2019 from www.politico.eu/article/steve-bannon-italy-europe-the-movement-experiment-will-change-global-politics/

Boyer, R., & Saillard, Y. (2002). *Regulation Theory: The State of the Art*. London; New York: Routledge.

Callenbach, E. (1990). *Ecotopia: The Notebooks and Reports of William Weston* (Bantam trade paperback ed.). New York: Bantam Books.

Carson, K. (2010). *The Homebrew Industrial Revolution: A Low-Overhead Manifesto* (Book Surge).

Carlson, T. (2019). The Truth about the Koch Brothers and the GOP. Retrieved July 1, 2019 from www.youtube.com/watch?v=8IgekVVqwkk

Dreher, R. (2017). *The Benedict Option: A Strategy for Christians in a Post-Christian Nation.* New York: Sentinel.

Elias, N. (2008). *The Established and the Outsiders.* The Collected Works of Norbert Elias, Dublin: UCD Press.

Elias, N. (2010 [1991]). *The Society of Individuals*, reprinted in R. van Krieken (ed.) *The Collected Works of Norbert Elias*, vol. 10, Dublin: UCD Press.

Elias, N. (2012a [1978]). *What is Sociology?* The Collected Works of Norbert Elias, Dublin: UCD Press.

Elias, N. (2012b [1939]). *On the Process of Civilization: Sociogenetic and Psychogenetic Investigations*, trans. E. Jephcott, reprinted in S. Mennell, E. Dunning, J. Goudsblom and R. Kilminster (eds.), *The Collected Works of Norbert Elias*, Dublin: UCD Press.

Farrell, W., & Gray, J. (2018). *The Boy Crisis: Why our Boys are Struggling and What we can do about It.* Dallas, TX: BenBella Books.

Fox, R. (2011). *The Tribal Imagination: Civilization and the Savage Mind.* Cambridge, MA: Harvard University Press.

Gellner, E. (1983). *Nations and Nationalism.* Cornell University Press.

Giddens, A. (1991). *The Consequences of Modernity.* Stanford: Stanford University Press.

Godwin, W., & Kramnick, I. (1976 [1793]). *Enquiry Concerning Political Justice: And Its Influence on Modern Morals and Happiness* (Pelican classics). Harmondsworth, Eng; Baltimore: Penguin Books.

Goodwin, M. (2018). "The national populist moment." *New Statesman, 147*(5439): 24–27.

Gorz, A. (1985). *Paths to Paradise: On the Liberation from Work.* London: Pluto.

Gorz, A. (2010). *Ecologica* (French list). London; New York: Seagull Books.

Hayek, F. (1945). "The use of knowledge in society." *The American Economic Review, 35*(4): 519–530.

Hazoy, Y. (2018). *The Virtue of Nationalism.* New York: Basic Books.

Hirst, P. (2001). *Associative Democracy: The Real Third Way.* London: Frank Cass.

Hobsbawm, E., & Ranger, T. (1983). *The Invention of Tradition.* New York: Cambridge University Press.

Homer-Dixon, T. (2007). *The Upside of Down: Catastrophe, Creativity and the Renewal of Civilization.* Toronto: Vintage Canada.

Ingold, T. (2000). *The Perception of the Environment. Essays in Livelihood, Dwelling and Skill.* London: Routledge.

Kauffman, S. (1995). *At Home in the Universe: The Search for Laws of Self-organization and Complexity.* New York: Oxford University Press.

Kolakowski, L. (1978). *Main Currents of Marxism: Its Rise, Growth, and Dissolution.* Oxford: Clarendon Press.

Kolozi, P. (2017). *Conservatives Against Capitalism. From the Industrial Revolution to Globalization.* New York: Columbia University Press.

Kish, K. (2018). *Ecological Economics 2.0: Ecological Economic Development Goals.* UWSpace.

Kish, K., Hawreliak, J., & Quilley, S. (2016). "Finding an alternate route: Towards open, eco-cyclical, and distributed production." *Journal of Peer Production, 9*(9, September).

Retrieved from http://peerproduction.net/issues/issue-9-alternative-internets/pe er-reviewed-papers/finding-an-alternate-route-towards-open-eco-cyclical-and-dist ributed-production/.

Kish, K., & Quilley, S. (2017). "Wicked dilemmas of scale and complexity in the politics of degrowth." *Ecological Economics, 142*: 306–317.

Lanz, T. (2008). *Beyond Capitalism & Socialism a New Statement of an Old Ideal: A Twenty-first Century Apologia for Social and Economic Sanity.* Norfolk, VA: Light in the Darkness Publications.

Lipietz, A. (1995). *Green Hopes: The Future of Political Ecology.* Cambridge, MA: Polity Press.

Loyal, S., & Quilley, S. (2017). "The particularity of the universal: Critical reflections on Bourdieu's theory of symbolic power and the state." *Theory and Society, 46*(5): 429–462.

MacIntyre, A. (2016). *Ethics in the Conflicts of Modernity: An Essay on Desire, Practical Reasoning, and Narrative.* New York: Cambridge University Press.

Meadows, D. (2008). *Thinking in Systems: A Primer.* London; Sterling, VA: Earthscan.

Médaille, J. C. (2010). *Toward a Truly Free Market: A Distributist Perspective on the Role of Government, Taxes, Health Care, Deficits, and More.* Intercollegiate Studies Institute.

Milbank, J., & Pabst, A. (2016). *The Politics of Virtue: Post-liberalism and the Human Future* (Future perfect: images of the time to come in philosophy, politics and cultural studies). Lanham, Maryland: Rowman & Littlefield International.

Morgan, T. (2018a). "#133: An American hypothesis. Is Donald Trump the first 'economic realist'." Retrieved October 1, 2018 from https://surplusenergyeconomics .wordpress.com/2018/08/24/133-an-american-hypothesis/

Morgan, T. (2018b). "#136: The Challenge for Government. The Politics of Declining prosperity." in *Surplus Energy Economics blog.* Retrieved October 6, 2018 from https:// surplusenergyeconomics.wordpress.com October 10, 2018.

Nozick, R., (2013 [1974]). *Anarchy, State, and Utopia.* New York: Basic Books.

Ong, W. (1991). *Orality and Literacy the Technologizing of the Word.* London: Routledge.

Odum, H. (2007). *Environment, Power, and Society for the Twenty-first Century: The Hierarchy of Energy.* New York: Columbia University Press.

Ophuls, W. (1977). *Ecology and the Politics of Scarcity: Prologue to a Political Theory of the Steady State.* San Francisco, CA: W.H. Freeman.

Ophuls, W. (2011). *Plato's Revenge: Politics in the Age of Ecology.* Cambridge, MA: MIT Press.

Orwell, G. (1962). *The Lion and the Unicorn; Socialism and the English Genius.* London, Secker and Warburg.

Pepper, D. (1993). *Eco-socialism: From Deep Ecology to Social Justice.* New York: Routledge.

Polanyi, K. (1944). *The Great Transformation: The Political and Economic Origins of Our Time.* Beacon Press.

Polanyi, K. (1957). "The economy as instituted process." in K. Polanyi, C. M. Arensberg, & H. W. Pearson (eds.). *Trade and Market in the Early Empires,* (pp. 243–270), New York: Free Press.

Quilley, S. (2011). "Entropy, the anthroposphere and the ecology of civilization: An essay on the problem of 'Liberalism in One Village' in the long view." *The Sociological Review, 59*(June): 65–90. doi: 10.1111/j.1467-954X.2011.01979.x.

Quilley, S. (2012). "System innovation and a new 'Great Transformation': Re-embedding economic life in the context of 'De-Growth.'" *Journal of Social Entrepreneurship, 3*(2): 206–229.

Quilley, S. (2013). "De-growth is not a liberal agenda: Relocalisation and the limits to low energy cosmopolitanism." *Environmental Values, 22*(2): 261–285.

Quilley, S. (2014). "Resilience through relocalization: Ecocultures of transition." in S. B. hm, Z. P. Bharucha, & J. Pretty (eds.), *Ecocultures: Blueprints for Sustainable Communities*, (chapter 12), London: Routledge.

Quilley, S. (2017). "Navigating the Anthropocene: Environmental politics and complexity in an era of limits." in Peter A. Victor, Brett Dolter (eds.), *Handbook on Growth and Sustainability*, (pp. 439–470), Cheltenham: Edward Elgar.

Scruton, R. (1984). *The Meaning of Conservatism* (2nd ed.). London: Macmillan.

Scruton, R. (2015). *How to be a Conservative*. London: Bloomsbury.

Tainter, J. A. (1988). *The Collapse of Complex Societies*. Cambridge, Cambridgeshire; New York: Cambridge University Press.

Tait, J. (2019). "Will Conservatives Abandon the Free Market." *National Interest*. April 18th. Retrieved October 24, 2019 from https://nationalinterest.org/feature/will-conservatives-abandon-free-market-53172

Thompson, E. (1991). *Customs in Common*. London: Merlin Press.

Wainwright, H. (1994). *Arguments for a New Left: Answering the Free-market Right*. Oxford, UK; Cambridge, MA: Blackwell.

Weiner, M. (2013). *The Rule of the Clan: What an Ancient form of Social Organization Reveals about the Future of Individual Freedom* (First ed.). New York: Farrar, Straus and Giroux.

Zywert, K. (2017). "Human health and social-ecological systems change: Rethinking health in the Anthropocene." *The Anthropocene Review*, 4(3): 216–238.

Zywert, K., & Quilley, S. (2018). "Health systems in an era of biophysical limits: The wicked dilemmas of modernity." *Social Theory & Health*, 16(2): 188–207.

3

NATIONS AND NATIONALISM IN THE ANTHROPOCENE

Steven J. Mock

It is a familiar cartoon; you've probably seen it on social media or a colleague's office door. It depicts a conference presentation in which the implications of measures to combat climate change are projected on a screen: "energy independence, preserve rainforests, sustainability, green jobs, liveable cities, renewables, clean water/air, healthy children, etc., etc." But an angry denier stands up from the back of the room and counters, "what if it's a big hoax and we create a better world for nothing?"[1]

The appeal of this cartoon lies in how succinctly it captures the bewilderment many of us feel when confronted with a denial of, or indifference to, the threat posed by climate change. Even if we concede for the sake of argument to scepticism and uncertainty about the science involved in assessing the gravity and source of the problem, many of the straightforward technical solutions devised to mitigate it have no apparent downside. They appeal to traditional values like frugality, purity, authenticity, and community as much as to liberal ones like social equality and protection from harm.[2] One would think that building an environmentally sustainable society should therefore be seen as a win–win proposition for all involved, but it is not. Instead, policies that contribute to environmental goals – even simple and innocuous ones like efficiency standards for light bulbs or time-of-use smart meters – mobilize active and vehement public opposition, mostly concentrated on the political right and expressed in terms that invoke notions of personal freedom (i.e. "I have a right to use whatever light bulb I want") and individual autonomy ("I don't want the government tracking my energy use"). Even those of us with sympathy over how wrenching the transition from a fossil fuel economy might be for some sectors of society struggle to understand why opposing bicycle lanes is a hill anyone would want to die on. Put simply: why should it be hard to get conservatives behind an agenda of conservation?

Several possible explanations present themselves, all offering insight, but none of which are entirely adequate. It is popular in leftist circles to blame the lobbying and public relations efforts of powerful corporate interests, primarily the fossil fuel industry. Certainly, the malign influence of money in politics, not excluding grassroots politics, is a genuine issue. But taken too far this explanation can tend toward a conspiracy theory that denies agency to masses of people whose beliefs are otherwise difficult to comprehend, assuming instead that money can simply buy their values and convince them to work against their interests. Public relations campaigns can have a dramatic effect on political mobilization, but only if they are successful in representing positions according to values and frames of reference that their target audiences already understand. This is evident in the fact that corporate and right-populist interests do not reliably coincide over issues related to environmental sustainability. Populist politicians have no difficulty mobilizing consumers and industry workers against the prospect of a carbon tax that even the likes of ExxonMobil supports.[3]

"Ideological bundling" suggests a partial answer to this quandary. Social Identity Theory (Tajfel and Turner 1986) and System Justification Theory (Jost, Banaji, and Nosek 2004) explain how the beliefs people adopt can derive not just from their own values, but also from the motivation to differentiate their identity group from outsiders and to legitimate the prevailing social order they rely on for existential security. The natural human tendency to frame political conflict in dichotomous left–right or liberal–conservative terms, exacerbated by partisan politics in which parties must construct platforms of positions over myriad issues that distinguish them from each other, leads to the clustering of beliefs into ideological bundles that consequently co-evolve in opposition to one another (Jost et al. 2013:314). As parties on the left of the political spectrum are increasingly preoccupied with environmental issues and concern for climate change, those who identify as conservative gravitate to opposing views in response. A tendency exacerbated by initiatives such as the Leap Manifesto in Canada or the Green New Deal in the United States that bind environmental issues to a socially liberal agenda and leftist economic policy at odds with the values and political identities of significant sectors of the population who might otherwise be persuaded to support them. These theories provide an effective framework for analyzing political polarization, but they leave open the question of how this particular co-evolution came about. Bundles of beliefs and values don't just cohere randomly. They must make some semblance of sense to the person holding them. What makes opposition to environmental sustainability coherent with modern conservatism?

Nationalism and the global growth economy

Understanding this evolution requires an understanding of the relationship between modern political identities and concept of the nation. In current popular discourse nationalism tends to be associated with reactionary, extreme-right ideologies. Since at least the Second World War and the horrors of fascism, the term "nationalism" evokes a narrow, inward-looking, exclusive, and aggressive worldview.[4] But this was not always the case. Between the French Revolution and the First World War, as monarchies and empires gave way to democracies and nation-states, nationalism was a liberal if not radical ideology; a movement that sought an end to ascribed classes and ethnic enclaves, against conservatives of the time who defended the inegalitarian hierarchies of the status quo. It stood for social mobility for the youth of the lower classes, and for either assimilation or self-determination for marginalized ethnic communities culturally or geo-graphically distant from imperial centres.

If the term in its current use tends to refer to only the most excessive and dys-functional manifestations of the ideology, it is because its progressive elements proved entirely victorious, melting into the unexamined background of politi-cal discourse as the nation-state became the fundamental unit of global political organization. As the original modernist theorist of nationalism Elie Kedourie put it at the outset of his treatise on the subject,

> not the least triumph of this doctrine is that such propositions have become accepted and are thought to be self-evident, that the very word nation has been endowed by nationalism with a meaning and a resonance which until the end of the eighteenth century it was far from having.
>
> *(Kedourie 1960:1)*

This happened not because these values and structures were morally superior, though we may judge them to be so. But because they led to – or at least were co-dependent upon – another ideational construct we have come to take for granted in the modern world, enabling a more functional and effective way of organizing the economy that could overwhelmingly out-compete societies organized along traditional lines: economic growth.

Nations, as subdivisions of the globe, may appear today as alternatives and challenges to a globalized economy. But the spread of "the nation" as the domi-nant mode of socio-political organisation was and is a global phenomenon. Long before the EU, the WTO, or the international corporation, cultures and polities worldwide were confronted with the same need to reorganize their societies in conformity with the norms that defined the modern nation-state in order to remain relevant and participate as fully functional units of the international system and economy. Modernist theorists of nations and nationalism have explained the intrinsic connection between this transformation and the "great transformation" that brought about industrial capitalist modernity.[5] I will draw primarily on the most systematic general theory of this kind on the origins of nations and national-ism; that of Ernest Gellner, who specifically modelled how the nation evolved as the form of social organization best suited to managing a growth economy.

The imperative of economic growth demands the expansion of the economy and an increase in its complexity, as the absorption of larger amounts of territory and larger numbers of people leads to greater specialization and broader networks of exchange and interaction. This necessitated two interrelated developments. The first was an ethic of meritocracy. Pre-modern agrarian societies that valued stability and continuity over progress and the maximization of wealth did not require individuals to excel in their occupations, merely fulfil them competently. Each role, including roles of leadership, required only a limited set of skills. Therefore the only criterion that a system for filling roles had to meet was the minimization of social strife. Heredity was as good as any other in that respect. One's position and life chances were determined by one's birth, and this system was reinforced by ethical and religious codes of the kind reflected in Plato's *Republic*, where morality was defined as all sectors of society accepting and working diligently in the assigned tasks of their ascribed stations. But a modern economy demands the maximization of human resources, requiring that the best people be placed in the right roles. You cannot fill the role of a company CEO the same way as you would that of a medieval Duke. Growth requires a system that can accommodate the upheavals of "creative destruction," whereby entrepreneurs from any sector of society are free to develop innovations that improve upon and replace existing technologies, enjoying temporary market dominance as a reward (Booth 2004:26; Victor 2008:34). Social mobility, in place of an ascriptive hierarchy of stations, must therefore be understood as a positive rather than a negative value. This in turn demands the embrace of the hitherto radical notion of baseline human equality – the concept that all people are in principle equal, at least in terms of their rights and opportunities at birth – so that the system can accommodate the replacement of less-qualified individuals in higher stations with more qualified individuals from lower ones with a minimum of social friction.

But even the most egalitarian society has boundaries; a centre and a periphery, with insiders and outsiders. Giving power to the people presupposes a shared understanding of who 'the people' are. This leads to the second necessary development of modernity: the transformation of culture into the principal signifier of political identity. In traditional agrarian societies culture could be taken for granted, as most transactions were confined to local communities. Culture therefore varied greatly within a given economic zone, with cultural difference – often characterized by different languages, religions, and myths of descent – serving to distinguish between, and thereby fortify, identities based on locality, social class, and economic niche. But economies driven by the growth principle must be prepared to constantly expand their reach beyond local face-to-face communities, while the need for specialization and mobility requires that all individuals functioning in the economy share common language, norms, and basic skills such as literacy, allowing them to interact irrespective of context and to change roles quickly to adapt to constant innovation. It is in this way that the nation came to prominence as the dominant form of political organisation, where signifiers of culture serve a unifying rather than distinguishing function.

As economic migrations occur, whether across territories due to uneven industrial development, or simply from the countryside to the city as a consequence of urbanization, the populations affected – whatever their pre-existing self-conceptions of identity and social status – find themselves at a disadvantage to the extent that cultural differences affect their ability to communicate and adapt in the economy and society. The same effect can occur even without mass migration, but simply as a consequence of the increased reach of the centre to the periphery through denser networks of communication and control that arise from the development of the bureaucratic state. The low culture of the periphery – its language, dialect, religion, or customs – is seen as ignorant and regressive against the high culture of the centre, to the extent that its differences hinder seamless social and economic intercourse. Since culture is, by definition, learned behaviour, peripheral populations find themselves confronted with a choice: they can either mobilize to assert the dignity and claim autonomy on behalf of their differing culture, forming a new centre or "nation" of their own; or they can adapt to the culture of their changed environment in the interests of inclusion and advancement within that society. We tend to assume that humans are inclined to choose the former option, because they only become political subjects when they do. But in practice the latter option is most often the path of least resistance. Thus the processes of modernity serve to destroy more potential nations than they create, as innumerable "wild cultures" are digested by the relative few that amalgamate or transform into national cultures (Gellner 1983:43–52).[6] The common language and skills of the centre, as a necessary facet of a functioning economy, become the individual's ticket to full functionality within that economy, and therefore his most treasured possessions, crucial to a sense of identity and belonging. The assurance of economic and social security is no longer a factor of one's birth, but of the extent to which one is in possession of the language, norms, traits, and skills that amount to the nation's high culture. Therefore, to maximize the productive potential of its population, the state is tasked with fixing the terms of the national culture, and disseminating it as widely as possible through mass education to ensure seamless communication and mobility of its human resources.

This is a brief explanation as to how the principles we now take for granted as definitive of a modern nation – in-group equality, social mobility, citizenship, mass politics, literacy, public education, and shared language and culture[7] – came to prominence in direct relation to the industrial economy's drive to maximize growth. The point is that the very terms according to which we conceive of notions like liberty, as agency to define one's self and determine one's fate free from constraint, are framed within a social construct that serves the function of economic growth. It may seem odd to equate liberty with what would appear to be its opposite: coerced conformity to the norms of a group such as a nation. But viewed in this light, it becomes clear why notions like "freedom" are so often conflated with a sense of belonging to one's nation; because in many ways these principles are indeed a gift of the group. Adhering to the norms of the group, to its common culture, is what allows for social mobility, a precondition to the free choice of one's role in the economy, providing a framework to recognize one's own dignity and equality vis-à-vis other group members. In the great transformation to modernity, the nation became the container for hitherto radical notions of liberty that the next generation of conservatives set out to conserve against any force that would contest them, be it communism, globalization, or environmentalism.

The degrowth of nations

It is here that the modern construct of the nation runs up against the problems we confront in the Anthropocene. To begin with, any norms that demand universal consent and enforcement to be effective are an anathema to the principle of national self-determination. As Gellner noted when he first posed his theory of the nation, a unified global society would not appear from the point of view of citizens of existing nations to be one of universal peace and harmony. It would amount to a global form of apartheid, where those distant from whatever was deemed to be the global high culture would be relegated to permanent subordinate status with no agency to construct a new political centre of their own (Gellner 1964:178). It is for this reason that at least some conservatives will resist any innovation, however trivial, that draws from an environmental agenda with global purpose at the expense of local or national norms and authorities. What liberal cosmopolitans close to the nascent global society view as mindlessly reactionary, anti-environmental behaviour is in fact entirely understandable for people for whom possession of the culture at the centre of their nation is the basis of their otherwise precarious economic security; for whom any loss of familiarity with that culture therefore presents as a threat to their mobility in society.

But the threat to the nation posed by the Anthropocene goes deeper than that. For if the growth economy is the underlying foundation of the modern nation, what happens when we recognize the limits of growth; when global society must embrace an ethic of limited growth if not degrowth? Can the nation survive? And, if not, what might we stand to lose along with it? Gellner speculated in the conclusion to *Nations and Nationalism* that if economic growth ceased to be a value, the nation could indeed cease to function as a construct:

> Much of our argument did hinge on the implications of continuing commitment to global economic growth, and hence to innovation and occupational change; it also pre-supposed the persistence of a society based on the promise of affluence and on generalized Danegeld. These assumptions, though valid now, cannot be expected to remain so indefinitely ... Our culturally homogeneous, mobile and, in its middle strata, fairly unstructured society may well not last forever, even if we disregard the possibility of cataclysms; and when this kind of society no longer prevails, then what

we have presented as the social bases of nationalism will be profoundly modified.

(Gellner 1983:113)

At the time, he did not need to develop this line of thought any further, concluding simply that, "the age of wealth-saturation for mankind at large still seems fairly distant, and so the issue does not affect us too urgently at present ... that is not something which will be visible in our lifetimes." But it may benefit us to revisit this speculation. For while the question did not prove urgent in Gellner's lifetime, when the sources and sinks for increased global energy throughput still appeared unlimited, it may well be in ours.

The notion that economies must continue to grow in order to be healthy, and that growth is therefore an uncontested boon to an economy, may be the only principle universally agreed upon across all sectors of our otherwise plural and anarchic international system. Commitment to growth as a policy priority, despite differences over the means by which it is best achieved, encompasses all types of state government, as well as international organisations such as the UN that promote social development, interstate alliances such as NATO that aim to enhance military security, as well as being explicit in the mandate of global institutions like the World Bank, IMF, and WTO (Purdey 2010). But as with the nation, this is a novel principle, however natural and self-evident it is taken to be in our time. Stability rather than growth was the ideal of pre-national agrarian societies. The notion that a functional society should perpetually increase depends on a decidedly modern conception of progress; a point that is acknowledged in both nationalism theory and sustainable development literature (see, for example, Gellner 1983:22, Victor 2008:4–9, also Purdey 2010:47–49, and Friedman 2005:22). But as we contemplate responsible environmental stewardship in the Anthropocene, our society will have to reassess this uncritical commitment to growth. Taking aside the basic thermodynamic principle that exponential growth cannot proceed indefinitely in a finite space, such as the closed system of our biosphere (Daly 1996:57–60, Homer-Dixon 2006:36–42), no social system, however natural or permanent it appeared at its zenith, has lasted forever. The history of human civilization is not a narrative of continuous progress, but cycles of increasing complexity followed by collapse (Tainter 1990). Even classical economics recognizes such a thing as detrimental growth: past an equilibrium where the marginal cost of a unit of increase is greater than the marginal return; a point we may already have passed in relation to our global energy throughput, as the increasing costs in environmental degradation outweigh diminishing returns in human comfort, security, and wellbeing (Daly 1996, 2014).

The problem is that even if reaching a limit to growth is inevitable on a physical, historical, and economic level, the ideational transition to a growth-oriented national society is and has always been one-way only. There is no going back, as the benefits it offers are so great as to make reversion to earlier forms seem unthinkable.[8] This is true both in a material sense, as national societies mobilize their human resources more effectively to out-produce all others, thereby affording them the ability to dominate the commons and offer their members a superior quality of life; and a moral one, as they activate notions of individual liberty that citizens are loath to surrender. Key here is Gellner's idea of modern society running on "Danegeld," an allusion to a period in Britain's ancient history when the Vikings were paid on a regular basis not to pillage the islands. In the modern equivalent, the masses are bribed not to overthrow the system through a level of comfort and security unheard of in previous eras to all but the uppermost echelons of society (Gellner 1983).[9] But this can function only in a growth economy, as the modern Danegeld necessary to keep industrial capitalist society stable is paid not solely in the currency of material abundance. It is also paid in the promise of freedom, embodied in the prospect of upward social mobility theoretically available to everyone, as well as to the progeny that will live after them, guaranteed by the nation that will, in principle, live in perpetuity. That promise cannot be maintained, even as an ideal, without growth. And so long as that promise is perceived to be embedded in the social contract, continued growth of the society as a whole will be an inevitable result. In any society where it is a value to aspire to better oneself socially, individuals are compelled to seek more than would be required simply to secure their own material satisfaction. They are compelled to seek excess wealth to secure better access to resources that will further their position in the competition for socio-economic roles, such as education, health, beneficial social connections, and so on.

One might object that such a system of competition does not make for a great deal of social mobility in practice. With the exception of brief periods of unsustainably overheated growth when large populations were able to collectively enhance their status, the norm has been relatively stable and de facto hereditary socio-economic clusters. But while social mobility may be the exception rather than the rule in practice, collective adherence to the principle has a dramatic effect on the trajectory of society as a whole. So long as there is perceived to be open competition for roles, in absence of norms ascribing fixed status, all members of society, whatever their status, must continually maximize their wealth and productivity simply to maintain status in the face of competition from lower strata. The aggregate result is the continuous expansion of society, as every productive member generates a surplus, and as the amount of education, wealth, and resources needed by all members of society to maintain their standing continually increases.

Aspirations and fears of this kind become even more acute when people contemplate prospects for their children in a society where their destinies will be determined by merit rather than birth. With the exception of only a handful of entrepreneurs, most adults accept by midlife that any major change in their economic or social circumstances through their own initiative within their own lifetimes is unlikely. But the circumstances of families can change enormously from one generation to the next, particularly given the right environment, investment, and attention, and this provides a powerful motivation shaping social and political attitudes (Friedman 2005:93). In the modern world where it is not taken for granted that children will inherit the stations of their parents, parents will be driven to acquire not just wealth sufficient to maintain and pass on their own livelihood, but surplus wealth to ensure maximum autonomy for their children, providing them with the means to freely choose their own destinies and raise their station.[10] A society where status and security means freedom of choice necessitates massive and constantly competitive levels of investment well beyond what would be required to maintain a family in a steady state of material comfort and security. Hence such a society cannot function outside of a growth economy. What's more, people will demand that their nation and its guardian state continue to grow relative to others so that it may remain competitive in its ability to fulfil its duty of providing an environment in which future generations can continue to maximize their mobility and security. These concerns are often explicit in the rhetoric used to campaign against government regulation and investment in the interests of environmental protection: the fear that it will lead to a command economy and debt load that that will weaken the nation relative to others, frustrating mobility and impoverishing future generations.

The growth imperative is not an elite-driven phenomenon, imposed from the top down – it pervades national society. To take one example, work-time reduction has been put forward as a policy for dealing with the stresses of unemployment consistent with the goal of transitioning to a non-growth economy without exacerbating wealth inequality (Victor 2008:211–214; see also Levy 2017). But while historically, work-hours reduction had been a signature goal and achievement of the labour movement, that tendency has shifted decisively in the post-war period to the point where policy experiments in work-time reduction attempted in recent decades have met their stiffest initial resistance from organized labour. Instead, political parties of both the left and the right reliably mobilize voters around platforms of job creation through economic growth. If it were to become widespread policy for companies to deal with unemployment by cutting salaries by 20%, but hiring 20% more workers for a 20% shorter work-week, objections would more likely come from workers and unions than from government, management, or from any wealthy elite.[11] It would be seen as threatening workers' opportunities to better their social status and that of their families through consumption, thus hindering their ability to participate fully in national life (see Ziegler, this volume).

It is not only a value in modern capitalist society to seek to better oneself and one's family; it is considered a fundamental human right. It is difficult to envision how this could continue in a steady-state economy, so long as the whole of the population continues to produce and consume with an aim to bettering their social station. In a steady-state society the principle of social mobility is not merely non-functional, it is positively subversive, as it was in pre-modern agrarian society, where social stability demanded that the son be content to inherit what his father had to offer, secure in the knowledge that accepting his station was what God's plan required of him. And flowing from social mobility is a host of positive values that we do not commonly associate with either capitalism or with nationalism: equality, literacy, education, civic culture, citizenship, civil society, mass political participation, and so on. Reverse-engineering Gellner's theory, we see how many principles crucial to our conception of liberty are dependent on the construct of the nation, which in turn relies on the positive valuation of economic growth for its functionality.

What then would be the fate of these values in a society where the engine driving the system, economic growth, was rendered impossible or undesirable? Too often, when we talk about the need to curtail our consumption culture in the name of sustainability, the image this conjures is one of SUVs, air travel, box stores, technical gadgets, and other luxuries important to our current way of life, but that we could imagine giving up without much insult to our basic morals. Rarely do we consider the environmental load of more intangible yet fundamental social goods as, for example, democracy, gender equality (explored further by Kish in this volume), or freedom from arbitrary violence. Education serves as an illustrative example. Pre-modern agrarian societies required little time and energy to train individuals in only whatever skills they needed to fulfil the roles assigned to them at birth. Literacy was rarely one of them. Mass public education, in contrast, requires massive investment on the part of modern societies. By way of illustration, it is estimated it takes the energy of 80 barrels of oil to provide an undergraduate education at an American college (Hall and Kiltgaard 2012:36), to say nothing of the energy put into mass secondary and even primary education. Most of this energy ends up wasted; how much of what we learn in school do we even remember, let alone use later in life? Such investment makes sense only in a dynamic growth economy that must produce cohorts of citizens with a vast repertoire of nearly identical skills such that they are capable of rapidly switching roles in the economy, yet that lives under a veil of ignorance as to how they may wish to or be required to do so in the course of their lifetimes. It makes macroeconomic sense only in a meritocratic society that benefits from allowing as many of its members as possible choose the roles they best excel in and find most satisfying, and therefore from the time and effort it takes to present them with as diverse an array of options as possible before they are required to choose. In a society that is not driven toward continuous expansion, how long will it be before the realization dawns that collective wellbeing is better served by conserving that energy and once again providing most members with only the training they need to fill ascribed roles? And how, if not through mature informed consent, will those roles be assigned?

Giving the Tea Party their due

It is not my intention to suggest that when economic growth slows or stops we will revert to a traditional agrarian order with barons and serfs. Too much has changed during these brief centuries of energy abundance. Technologies have been invented that cannot be uninvented,[12] but also values have taken hold that cannot be easily overturned. We have come to embrace modern conceptions of freedom in their own right, and most would consider this a good thing. And while our embrace of values such as social mobility causes us to maintain our commitment to the engine – economic growth – on which they rely, even if this engine is removed entirely, and the principle ceases to function, our commitment to these values may ensure that we find some way to preserve them.

But it is incumbent on us to confront this as a dilemma, not just for the purpose of socially engineering an Anthropocene we can live with that will maintain the values we wish to preserve; but also if we expect to enlist the support in this endeavour of those, like the guy in the cartoon, who struggle to see this vision of the future as an unambiguously "better world," instead recognizing moral trade-offs implicit in its construction. For far from being the mindless obstructionist he seems, he may in fact be drawing from moral intuitions others of us lack, pointing to ways that efforts at environmental preservation, even innocuous ones with seemingly beneficial side-effects, conceal threats to notions of liberty crucial to modern identities. For while I would not venture to pronounce definitively on what alternatives to the norms of the modern nation might be suitable to the functioning of a futuristic steady-state society, the possibilities one might speculate as consistent with this theory are sufficiently alarming.

One of them would be the de-emancipation of women. Transforming an arbitrary half of the population into unpaid caregivers would instantaneously reduce the production capacity of society and the level of competition-related consumption. Another would be racism. For while it would be unrealistic, given the intervening developments of modernity, to expect that society would revert to a pre-modern agrarian system of hereditary stations legitimated by divine mandate, a secular-rationalist discourse of pseudo-scientific categorisation could be used to justify dividing superior and subordinate groups according to arbitrary superficial traits, thus determining socio-economic role, access to education and culture, and the distribution of fixed resources in a manner that curtailed competition.

Whatever moral arguments can be made against racism and sexism in terms of justice, protection from harm, and the value of tolerance, I would argue that one of the reasons they remain among the most universally abject ideas in modern society is because they are dysfunctional to such a society, an unproductive hindrance to the optimal mobilization of human resources further to the goal of maximizing growth. Which is not to say that racist and sexist attitudes and systemic discrimination do not persist in modern society, but those who think this to be a good thing, who embrace this as a norm, are confined to an extremist fringe usually associated with other aspects of anti-modernity. Would it remain so if the principles of equality and mobility ceased to be functional as well as moral, but rather became a threat to society alongside the growth they encouraged?

The point is that anxiety in response to measures that promote environmental sustainability does not reflect mere irrational obstructionism. It may well be the product of an implicit awareness that the society in which we live, including positive values we have come to understand as the very definition of freedom, is indelibly linked to economic growth. Viewed in this light, modern conservatives resisting environmental initiatives do so in defence of egalitarian values and shared cultural norms; against radical greens whose agenda amounts, consciously or not, to an assault on these principles in the name of a top-down order of ascribed hierarchy, inimical to justice and morality as most members of modern societies understand them.

Understanding this connection between social mobility and economic growth, and how these in turn tie in to the construct of the nation, is crucial to grasping the emotional force that the growth imperative wields in the modern world, enabling it to persist even in the face of scientific arguments and empirical evidence that it is unnecessary and even detrimental to long-term ecological sustainability and human happiness. We must recognize that conceptual challenges to the growth economy are indeed existential challenges to the very core of modern notions of social stability, justice, and identity: constructs such as the nation, institutions such as democracy, and values such as human equality. So long as the prospect of growth is so intimately connected to our basic values and deepest anxieties, both the continuity of governments in democratic systems and the survival of regimes in non-democratic ones will depend on the level of growth they can maintain.[13] Under such circumstances it is difficult to envision, whatever the stakes, any credible political movement that could advocate, or government that could implement a program for transforming the national economy away from the growth principle, even if such an economy was able to maintain an acceptable standard of living for its members.

If indeed the nation is a phenomenon contingent on modernity, the question of whether it can survive a genuine transition into post-modernity, and what we stand to lose if it does not, has dramatic implications. And finding a way to preserve the values of modernity in a non-growth economy – demonstrating that it can be done – might be necessary to address the anxieties that stand in the way of mobilizing humanity for sustainable life in the Anthropocene.

Notes

1 Drawn by Joel Pett, it first appeared in USA Today in December 2009 just before the Copenhagen climate conference. https://www.kentucky.com/opinion/op-ed/article44162106.html
2 Roughly paraphrasing Jonathan Haidt's (2012) model of the distinct moral intuitions emphasized by conservatives and liberals, respectively.
3 At time of writing, Canada is experiencing its own "yellow-vests" protest movement, one of whose primary demands is opposition to a carbon tax, while a recent issue of the Economist (09/02/2019) noted in a special report on ExxonMobil's plan to ramp up fossil fuel production that, while this was a business decision consistent with current policies and market trends, the company formally advocates a carbon tax as a "transparent and fair way to limit emissions."
4 Though related terms like "patriotism" may be deployed to refer to more positive, constructive aspects of what is in essence the same ideology (see Wimmer 2019).

5 Karl Polanyi, who coined the term "the Great Transformation" (1944) could be considered the first of these, but subsequent elaborations that specifically focus on the development of nations and nationalism include Nairn 1981, Hechter 1975, Anderson 1991, and Gellner 1983.

6 Many elements of this process, minus their implications to national identity per se, are mirrored in Friedman's (2005:304–310) description of the modernization of developing countries.

7 Anthony Smith (1991:14) defines a nation as "a named human population sharing an historic territory, common myths and historical memories, a mass public culture, a common economy and common legal rights and duties for all members."

8 As explored at length by Quilley in Chapter 2.

9 By way of illustration, Thomas Homer-Dixon (2006:83) calculated that a tank of petrol in an average car harnesses energy equivalent to two years of human manual labour. In effect, every functional member of our society can afford their own slaves. It has been estimated that an annual growth rate of 3–5% GDP is necessary to the long-term maintenance of social stability (Booth 2004:37–53).

10 Friedman (2005) gives special attention to American society as the exemplar and historical instigator of this global social change. The classlessness or "equality of condition" of American society was what most struck de Tocqueville as the basis of its productivity. "Everyone wants either to increase his own resources or provide fresh ones for his progeny" (quoted in 105–106). Opening the opportunity for advancement to all heightened the incentives prodding economic effort and, indeed, increased the costs of failing to engage in such effort. Given the specific challenges faced by the United States at the time, the entrepreneurial spirit engendered by social mobility was especially needed and was thus framed not just as an opportunity but as an obligation, for the individual as well as for the nation. "As a result, Americans have consistently interpreted freedom itself at least partly in terms of their dedication of energy toward economic ends and their ongoing success at that endeavour" (109).

11 Though this simple supposition would be complicated by issues of class and national culture. Where worktime reduction has been tried as an instrument of employment policy by several countries in Europe, it has often won over at least some well-paid professionals who found that the additional income was of less value to them than additional time spent with their families or on their leisure pursuits (see Kallis et al. 2013 for a critical analysis). One can imagine that it would be a harder sell among, say, factory workers in the United States, in circumstances where lost income would translate more directly into lost status and security, where the capitalist ethos is more firmly embedded in the national identity transcending class lines.

12 To begin with, this chapter has not addressed the argument prevalent in sustainable development literature that a steady-state economy could maintain qualitative growth in the form of technological innovation to enhance efficiency, without quantitative growth in the form of continuous increase in the energy throughput of the system. Regardless, this will necessitate voluntary acceptance of limits to the economy's growth potential, and imagination will still be required to envision a society capable of such nuanced distinctions, conditioning mass revulsion toward the mechanisms of quantitative growth while embracing the qualitative.

13 A correlation substantiated by Friedman 2005:320–325.

References

Anderson, B. (1991). *Imagined Communities: Reflections on the Origin and Spread of Nationalism*. New York: Verso.

Booth, D. E. (2004). *Hooked on Growth: Economic Additions and the Environment*. New York: Rowman & Littlefield Publishers Inc.

Daly, H. (1996). *Beyond Growth: The Economics of Sustainable Development*. Boston: Beacon Press.

Daly, H. (2014). *From Uneconomic Growth to a Steady-State Economy*. Cheltenham, UK and Northampton, MA, USA: Edward Elgar.

Friedman, B. M. (2005). *The Moral Consequences of Economic Growth*. New York: Knopf.

Gellner, E. (1964). *Thought and Change*. Chicago: University of Chicago Press.

Gellner, E. (1983). *Nations and Nationalism*. Oxford: Blackwell Publishers.

Haidt, J. (2012). *The Righteous Mind: Why Good People are Divided by Politics and Religion*. New York: Pantheon Books.

Hall, C., & Klitgaard, K. (2012). *Energy and the Wealth of Nations. Understanding the Biophysical Economy*. New York: Springer.

Hechter, M. (1975). *Internal Colonialism: The Celtic Fringe in British National Development, 1536–1966*. London: Routledge.

Homer-Dixon, T. (2006). *The Upside of Down: Catastrophe, Creativity and the Renewal of Civilization*. Toronto: A.A. Knopf Canada.

Jost, J., Banaji, M., & Nosek, B. A. (2004). "A decade of system justification theory: Accumulated evidence of conscious and unconscious bolstering of the status quo." *Political Psychology, 25*(6): 881–919.

Jost, J., Federico, C. M., & Napier, J. L. (2013). "Political ideologies and their social psychological functions." In M. Freeden (Ed.), *Oxford Handbook of Political Ideologies*. New York: Oxford University Press, pp. 32–250.

Kallis, G., Kalush, M., O'Flynn, H., Rossiter, J., & Ashford, N. (2013). "'Friday off': Reducing working hours in Europe." *Sustainability, 5*(4): 1545–1567.

Kedourie, E. (1960 [1994]). *Nationalism*. Oxford: Blackwell, 4th edition.

Levy, A. (2017). "Prometheus unwound: Shorter hours for sustainable degrowth." Chapter 14 in *Handbook on Growth and Sustainability*, Edward Elgar Publishing, pp. 303–325.

Nairn, T. (1981). *The Break-up of Britain: Crisis and Neo-Nationalism*. London: Verso.

Polanyi, K. (1944). *The Great Transformation: The Political and Economic Origins of Our Time*, 2nd edn (2001), Boston, MA: Beacon Press.

Purdey, S. J. (2010). *Economic Growth, the Environment and International Relations: The Growth Paradigm*. New York: Routledge.

Smith, A. D. (1991). *National Identity*. London: Penguin Books.

Tainter, J. (1990). *The Collapse of Complex Societies*. Cambridge: Cambridge University Press.

Tajfel, H., & Turner, J. C. (1986). "The social identity theory of intergroup behaviour." In S. Worchel & W. G. Austin (eds.), *Psychology of Intergroup Relations*. Chicago, IL: Nelson-Hall, pp. 7–24.

Victor, P. A. (2008). *Managing Without Growth: Slower by Design, Not Disaster*. Cheltenham: Edward Elgar.

Wimmer, A. (2019). "Why nationalism works (and why it isn't going away)." *Foreign Affairs, 98*(2): 27–34.

4

RECLAIMING FREEDOM THROUGH PREFIGURATIVE POLITICS

Kaitlin Kish

This chapter uses complexity theory to further explore the rigidity of the current system. In doing so, I argue that the current golden age of individual liberty and freedom has functioned as a distraction so that the system could be increasingly controlled to move in one single direction that increasingly benefits certain people. Throughout this process, societies have lost their ability to function as a complex adaptive system – instead, society is moving along a single trajectory that is either nearing the precipice of collapse or has already passed the threshold. Collapse of the system is inevitable, not least because of diminishing returns, but also because Western industrial capitalism has led to the normalization of order and structure in one of the most complex systems ever witnessed. This limits humanity's ability to establish a mutually enhancing relationship between humans, non-human species, and the Earth as cultural and ontological adaptation in such a rigid and monoculturous system is difficult, if not impossible.

I begin with a brief overview of relevant complex systems concepts. I relate these systems to our current predicament of capitalist modernity. In doing so, I argue that liberal notions of freedom have become increasingly irrelevant to the reality of our socio-ecological emergency. Under the stress of declining access to cheap energy and oil, political and social systems will be forced to open to greater pluralism in thinking about how we live. While Quilley, in this volume, argues for a more libertarian approach to life within this reality, I argue that a systems-based approach to liberty and freedom inherently means allowing for self-organization in a system – localized anarchy. I explore how this relates to anarchist thinking to empower localized prefigurative politics. I finish the chapter with a discussion of women and motherhood as an example of self-organizing prefigurative politics.

A brief introduction to complexity and resilience

This chapter is founded on complex systems philosophy and approaches from the field of socio-ecological resilience (Gunderson and Holling, 2001; Berkes, Colding, and Folke, 2000; Berkes and Folke, 2000; Holling, 1973), which look to describe how multiple elements interact overtime. The specific frameworks of resilience and panarchy are used to understand the ways systems transform and how properties of a system change over time. The properties of complex systems especially relevant for this paper are emergence and tipping points. Emergence is when a system exhibits properties that would not occur when parts of the system existed separately from one another. These properties only occur when the pieces of the system interact as a whole. A common example are groups of ants who spontaneously assemble themselves into tiny bridges to achieve a higher order need of the system. A tipping point is when the system passes a threshold and the system crumbles, such as the ant bridge becoming too long and collapsing due to instability. Once the collapse happens the system falls into a new state, or what systems theorists call a new "basin of attraction."

Tipping points, and system behaviour, are somewhat predictable in that they follow certain patterns. Socio-ecological scientists plot one overarching pattern by using the concept of "panarchy." Panarchy has two integral pieces to it: 1) the resilience framework and, 2) the adaptive cycle metaphor – I focus on the latter.

The adaptive cycle metaphor is characterized by a) a four-phase cycle, b) panarchy, and c) three distinct kinds of change. Holling and Gunderson argue that most socio-ecological systems follow a four-phase cycle:

1) exploitation (r); 2) conservation (K); 3) release (Ω); and 4) reorganization (α)

The first and second phases come from ecological theory in which "an ecosystem's r phase is dominated by colonizing species tolerant of environmental variation and the K phase, by species adapted to modulate such variation" (Gotts, 2007, p. 2). The third phase, a rapid phase such as a forest fire or insect outbreak that frees nutrients from biomass, is sometimes referred to as "creative destruction" because as the adaptive cycle moves into the release phase, power and resources that were once tightly consolidated within the dominant basin of attraction are freed up and made available for use by other actors, including pre-existing alternatives (Holling and Gunderson, 2002, p. 45). It is within this space of creative destruction that prefigured political ideologies can coalesce around an alternative system state, deepening the alternative basin (Marx, 1844; Schumpeter, 1942, 1947). In the fourth phase, "resilience and potential grow, connectedness falls, unpredictability peaks, and new systems entrants can establish themselves" (Gotts, 2007, p. 2).

The adaptive cycle is an element of "panarchy," Figure 4.1, which refers "to the framework for conceptualizing coupled human-environment systems" (Gotts, 2007, p. 1).

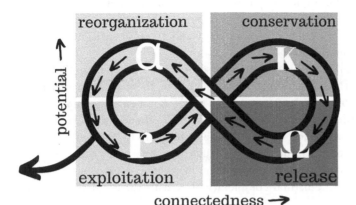

FIGURE 4.1   The adaptive cycle

While the adaptive cycle has received criticism in its application to ecosystems (Janssen et al., 2006), it is a useful way to conceptualize how change can happen in complex systems over time, particularly in its contribution to the metaphor of the gravitational landscapes (Figure 4.2), which helps to conceptualize the process of transition. This metaphor has been applied over the past two decades to ecological systems in the sub-discipline of resilience studies (Walker and Salt, 2006).

The gravitational landscapes metaphor focuses on three key concepts of a system: a) a system's resilience, b) a system's thresholds, and c) the "basin of attraction" in which the system lies. Within this field of study, resilience refers to a system's ability to adjust, rebound, or avoid crossing a threshold. If a system passes a critical threshold, it falls into a new "basin of attraction," with its own distinct structure and patterns of feedback. The more difficult it is to pass the critical threshold, the more resilient the system.

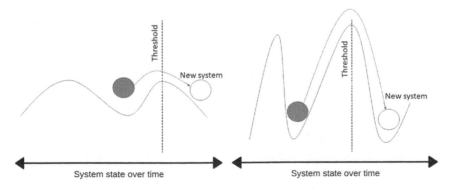

FIGURE 4.2   Basins of attraction over time with two different levels of resilience between system states

Gunderson and Holling identify three types of change that happens within panarchies: 1) incremental, smooth, and fairly predictable changes between the r and K phases; 2) abrupt change from K onward, and 3) transformational learning among panarchical levels.

This chapter deals primarily with the second kind of change (2002). In dynamical systems theory, within this kind of change, there are always surprises which make prediction of a system's outcome nearly, if not entirely, impossible. It therefore seems futile that we might try and plan for resilience of a very complex system over a very long period with multiple possible paths for the system to take. This chapter thus doesn't argue for a particular outcome in a system. Instead, I argue that we need to allow for exploration of creative approaches to system design that allow for the natural complexity of systems to re-emerge pre- and post-transition.

Socio-ecological complexity and the adaptive cycle: Modern capitalism on the brink

Panarchy and the adaptive cycle work as a mental map for how systems can change over time, but do complex socio-ecological systems follow a panarchic process?

Human history seems to follow a pattern of rise and fall of great civilizations, with collapse following a great climax in their triumph. As societies reach the peak of their scale and complexity, they require greater energy for their maintenance, to the point where they reach a point of declining marginal returns (Tainter, 1988). At that point, societies are in the K phase of the cycle. Then, new challenges emerge that require greater energy to solve and the society is too inflexible in its response to appropriate change, leading the society into the third phase of creative destruction. Global capitalist systems are certainly at the zenith of their complexity and seem to be showing a great inflexibility at changing to adapt to decreasing marginal returns and energy availability. For centuries, modern society has used cheap energy to develop greater and greater complexity and avoided collapses of sub systems, such as the financial system, by forestalling collapse. Boyd argues that this slight deviation from the adaptive system is due largely to the fact that human societies are led by their cultural models and beliefs systems – a unique characteristic of human systems. Thus, human systems may sometimes diverge from the regular flow of the panarchy cycle:

> Through such processes as cognitive dissonance human groups can reject new knowledge and experiences that challenge fundamental beliefs. This holds the possibility that the required change in belief systems will significantly lag the start of the release period, possibly severely impacting the ability of human society to effectively respond to its new circumstances.

(Boyd, 2016)

Instead, energy availability will have to inevitably force change in the system.

Conversely, Gotts argues that large national and transnational global and political systems will be the primary determinants of large-scale social systems transformation from K onward (2007). While Holling argued that the 1989 fall of the Berlin wall was indication of the global system entering the backloop Ω and α phases of an adaptive cycle, the subsequent economic growth suggests otherwise (Gotts, 2007). Such resilience of the capitalist system has made it seem impossible to track the planetary system to any convincing certainty on the adaptive cycle. However, Gotts says:

> Connectedness and certain kinds of potential are rising, and resilience is probably falling, suggesting that we are in a K phase. At the same time, continuing rapid innovation suggests an r phase, whereas the growing release of stored energy from fossil fuels, plus soil erosion, extinctions, and deforestation would seem to indicate Ω … Whether through war, full-scale ecological collapse, or a technological and/or socio-political revolution, it seems certain, as Holling (2004) notes, that radical global change is coming in this century, but this is evident even without the panarchical perspective.
>
> *(Gotts, 2007)*

Whether the system rigidity is enforced by cultural beliefs or the powerful resilience of transnational economic systems, there is a great deal of indication that modern capitalist society is in the third phase of the panarchy cycle. And, accepting that global human systems are governed by principles of the adaptive cycle, means accepting that contraction and radical reorganization are healthy periods of system transition. This stands in very deep contrast to contemporary economics which accepts that growth is the only healthy characteristic of a system. The adaptive system metaphor stands in strong defiance to neo-classical traditions that view the economy as self-regulating toward equilibrium. Instead, economies experience sudden shifts and periods of contraction, collapse, and reorganization.

When I refer to "collapse" I am pointing toward the decline or reorganization of the current variety of capitalism and globalization toward a new political economy. There are many physical system indications (Rockstrom et al., 2019), as well as theoretical indications that we may be on the eve of such reorganization:

> The machinery of government, entrusted with the maintenance of the existing order, continues to function, but at every turn of its deteriorated gears it slips and stops. Its working becomes more and more difficult, and the dissatisfaction caused by its defects grows continuously. Every day gives rise to a new demand. 'Reform this,' 'Reform that,' is heard from all sides … And yet all known that it is impossible to make things over, to remodel anything at all because everything is interrelated; everything

would have to be remade at once; and how can society be remodelled when it is divided into two openly hostile camps.

(Kropotkin, 2002, pp. 36–37)

Kropotkin goes on to ask how society can jumpstart a revolution. "The answer is easy" he says. *"Action*, the continuous action, ceaselessly renewed, of minorities brings about this transformation. Courage, devotion, the spirit of sacrifice, are as contagious as cowardice, submission, and panic" (2002, p. 38).

All forms of action are required; collective or individual, safe or daring, private or public for reorganization. Action is required to keep the spirit of revolution alive, "to propagate and find expression for dissatisfaction, to excite hatred against exploiters, to ridicule the government and expose its weakness, and above all and always, by actual example, to awaken courage and fan the spirit of revolt" (Kropotkin, 2002, p. 39). One possible path for such action is through a wide adoption of prefigurative politics.

The potential of prefigurative politics

Once the planetary system reaches, and passes, the critical threshold, there is a gradual or sudden breakdown of order. Ehrlich and Ehrlich argue that modern capitalist society is already in the process of a gradual breakdown, and thus the modern socio-economic system is now moving toward collapse and reorganization (2010). This is the peak moment for opportunity to implement change strategies for transformation – there is hope and potential beyond this "seemingly inevitable self-destruction brimming at the edge of too much complexity" (Erlic, 2016). This opportunity is taken up by what Kauffman calls "the adjacent possible" (1996). The adjacent possible

> is a kind of shadow future, hovering on the edges of the present state of things, a map of all the ways in which the present can reinvent itself … [the adjacent possible] captures both the limits and the creative potential of change and innovation.
>
> *(Smith, 2010)*

I explore the adjacent possible through one particularly inspiring, and easily accessible, area of opportunity: the theory of prefigurative politics. Prefigurative politics are those who establish alternative political or cultural systems that become opportunities for system transition. Multiple prefigurative political orientations, linked together by common goals, could come together as a strong alterative basin, as explored in Quilley's chapter on liberty in the long Anthropocene.

Theories of prefigurative politics have their roots in social opposition to the disruptive forces of industrialization and bureaucratization in early 20th-century Europe (Boggs, 1977). Between 1918 and 1920, Gramsci developed a theory of transition toward socialist democracy that identified factory councils in Italy as key elements of a radical political order that would take hold in the spaces opening up as a result of escalating crises in bourgeois society (Boggs, 1977). This conceptualization of the potential of prefigurative experiments to gain ground as crises accumulate in the dominant regime is now embedded within theories of socio-ecological systems transformation in resilience studies and social innovation (Westley et al., 2007). For social movements working toward radical change, prefigurative politics is a way to enact new patterns of social relations that can be imagined from within the current system, but that diverge too much from the mainstream to gain widespread traction under existing conditions (Breines, 1980).

We have seen the success of prefigurative politics in historical social movements where practitioners have created and/or embodied the ontologies and structures they envision for a transformed society. Most social movements in the 20th and 21st centuries, such as those for women's rights, the environment, peace, anti-globalization, anti-capitalism, economic equity, and indigenous rights have included prefigurative elements (della Porta et al., 2010; Graeber, 2009; Cornish et al., 2016). There are many examples of existing prefigurative groups, such as anarchist groups Abalone Alliance (anti-nuclear), Clamshell Alliance (anti-war), Food Not Bombs, as well as organizations such as lending libraries and community gardens. These groups are trying out new ways of life. Bioregionalism, ecovillages, and the permaculture movement are further examples of prefigurative communities that embody relational world-views and are characterized by embedded socio-ecological relationships (Lockyer and Veteto, 2013).

Breines argues that the crux of prefigurative politics lies in the substantial task for individuals to live the practice of their movement so that relationships and political forms of the desired society are already in action (1980). John Holloway argued that, for those seeking to fundamentally transform society, the solution is simple: "Refuse-and-create!" (Holloway, 2010, p. 50). The route to overthrowing capitalism, Holloway argues, "lies in the proliferation of small-scale rebellions against capitalist logic" (Young and Schwartz, 2012, p. 221) envisioned by "a multiplicity of interstitial movements" (Holloway, 2010, p. 11) all with the same unifying thread: to overcome the alienation characteristics of capitalist labour and replace it with work and activities that are fulfilling, voluntary, and socially useful (p. 198).

While a lot of current literature on prefigurative politics focuses on how activists should build social movements (Epstein, 1991; Polletta, 1999), the "original concept of prefigurative politics involves a politicization of everyday life," (Williams, 2017) so to capture the full spectrum of prefigurative politics we need to see changes in everyday life as radical acts of resistance – such as making, parenting, and being with family – which I return to at the end of this chapter. In Karp's chapter in this volume, he explores the role of activists – prefigurative practitioners should be included in the definition of an activist.

Freedom from a complex systems perspective

One challenge facing prefigurative activists is the rigidity of highly resilient systems. The conservation stage of capitalist modernity has created strong interconnections across global subsystems; local production has been replaced by globally integrated systems, local banks have been replaced by national banks, locally produced food has been replaced by multinational food systems relying on cash crops, and local currencies and systems of trade have been homogenized (i.e., the Euro). This is a much more efficient system, "but much of its resilience has been the price of that efficiency" (Erlic, 2016). The global system has becoming intimately tied to, and thus dependent on, ecological and resource niches. Additionally, capitalist modernity's tendency toward cultural monocultures is "much less resilient to challenges" (Erlic, 2016). For example, a decline in the value of the Euro has enormous cascading impacts across the world. According to panarchy, while many would characterize the last few hundred years as highly liberated with freedom for individuals, we have been stuck in a conservation stage – characterized by high rigidity in the system overall through monocultures and replacement of local systems with larger global systems. The only reason individual agents have had a notion of freedom is due to the process of disembedding which allowed for greater spatial and temporal flexibility in life, which at the same time contributed to the overall rigidity of the system (disembedded and "free" agents underpin the capitalist paradigm).

Thus, freedom of the individual feeds the rigidity of the overall system and the overall rigidity of the system necessitates more freely mobile individuals – it has created a strong feedback loop at the cost of the environment.

For example, one's freedom to consume further perpetuates the resilience of a capitalist system which, in turn, ensures the individual is stuck in a consumer/worker role. And to keep this system within that state, we have impositions of order from the top down. The Western political economy is constantly fighting against the natural chaos of the system to ensure individuals remain "free," while decreasing the freedom from living outside of the system. Those who want to live outside of bureaucratic orders that prevent chaos, are often shunned such as homeschoolers, preppers, homesteaders, ultra-religious groups – because they threaten the strong resilience of the current economic system – our liberty and freedom to live the life we choose is limited by how much it adheres to the overall definition of a "citizen" in society. Nations and corporations stand above the well-being of individual humans and nature for their own self-interest by imposing significant pressures of top-down control.

Western society has come to normalize order, control, centralization, and discipline as a necessity for the modern world. This restricts systemic tendency toward self-organization and desire for self-control. According to complexity theory, the least energetic way to generate order is by allowing for self-organization and emergent properties to take natural course without the need for outside control and structure. That is, to dismantle an overarching system that restricts an individual's role within the system, and instead allowing for someone to do what might feel more naturally compelling and significant within their localized context.

From this view, liberal notions of freedom will become increasingly out of touch with social and natural reality and that, under stress, the cultural/political system will open to greater pluralism in thinking about liberty, including a rediscovery of and building upon of 19th-century anarchist thinking. Normalization needs to shift to empowerment of the needs of individual communities, and we certainly see this happening more often through group such as Maker communities, adoption of alternative health systems, intentional communities, Transition Towns, and others. We are starting to see many people coalescing around what makes sense for their local context in the face of future uncertainty.

Erlic argues that beneficial policies will be those that focus on diversity and forestalling economic and social concentration (2016). "A diversity of worldviews would also be beneficial, and therefore a political system that provides space and support for divergent beliefs," (2016) which would allow for greater growth of prefigurative politics.

To do so, existing institutions do not need to be destroyed – as this would simply make room for new institutions. Rather, institutions should become profane. Newman argues that we need to start to think and act as if this top-down power no longer exists, live the prefigurative life. This is not just freedom, it is the "self-determination of owners invested in themselves, and through themselves, in others" (2016). If reductionist legal and social systems have created a static system reliant on institutions, it is nonadaptive in a time when it is increasingly important for it to be adaptive given the unpredictable time. It seems contradictory that the road to greater ordered freedom is through fewer codified rules regarding our roles and responsibilities; but the nature of complexity is order through chaos, as this creates room for the creativity of nature to produce localized rights, freedoms, and empathy as the local self-organizing system needs.

Such reorientations of thought can come from the most seemingly mundane areas of life. For example, in Western societies, capital, labour, and product have come to be somewhat synonymous. Unlimited and efficient labour is required to produce the highest number of products to achieve the greatest margin of capital – this is common sense in Western capitalist economics. However, Marx declared capital and product to be different, suggesting that product and production must be seized by society. Not only are they different, but they no longer serve an appropriate function for the well-being of society and "it becomes evident that the economic institutions which control production and exchange are far from giving to society the prosperity which they are supposed to guarantee; they produce precisely the opposite result" (Kropotkin, 2002, p. 36). Additionally, Tim Ingold incites the romanticism and spirt associated with making and producing:

> The draughtsman with her pencil, just like the carpenter with his saw, must feel where she is going, and must continually adjust her gestures so as to maintain alignment with a moving target. Moreover, as with the mountain path, the buzzard's flight or the tree root, the drawn line does not connect predetermined points in sequence but 'launches forth' from its tip leaving a trail behind it.
>
> *(Ingold, 2011, p. 99)*

There is a human-ness to production that has been stripped out to maintain a certain line of order in society. The Maker movement and citizen reclamation over production is one seemingly mundane prefigurative political action that encourages self-affirmation, creativity, spirit, and localized autonomy over self, life, and nature.

Resilience principles ask us to manage slow variables and feedbacks, to configure the system in different ways to make all variables in a system both independent and connected to provide different services. The only way to do this is by empowering the roles within different parts of the systems through radical liberation. The philosophical parallels between complexity and anarchism begin with this notion of self-organization as both emergent and better for the long-term resilience of a system. Embracing complexity inevitably leads one toward an acceptance of anarchy as current modes of liberty and freedom are upheld by incredible systems of hierarchical control (and a history of complex growth dynamics that have reaped significant damage on ecological systems). Now that hierarchical control is put under pressure and brought into question because of the precarious relationship of socio-ecological systems.

While anarchy is traditionally equated with "chaos," and usually associated with disorder, it is self-organization from the bottom-up:

> Indeed, anarchy is the condition for a further search for principles and rules and it reveals, moreover, as the conditions to any possible life and organization to be. In other words, it is the autonomy, the freedom, the independence that allows for higher or better horizons, dynamics and structures. Anarchy, it appears, is the very seed and proper name for freedom, autonomy and cooperation.
>
> *(Maldonado, 2016, p. 60)*

Localized systems without political rule could reduce socio-political instability and increase socio-ecological relationship especially in context where fighting for "land" isn't a thing anymore. By reducing significant bureaucratic overhead, we can also reduce energetic demands, produce local food systems, and reorient how individuals in local communities relate to one another. This is certainly not something that could be put into practice in any large scale – I'm certain a large city would not function without top-down systems. This is a prefigurative possibility for those in a position to create change – municipalities willing to go a different direction or small communities interested in establishing local resilience. For example, in Prince Edward Island there is a thriving community of local food, limited reliance on imports, sharing, gifting, bartering, trading, and livelihood economics initially fuelled by economic instability (Kish, 2018). Without top-down control, the community began to self-organize in an anarchist fashion to meet the needs of the community. Kropotkin, taking a cue from Darwin, argued for networked societies based on mutual aid, compassion, trust, and solidarity as grounds for a truly free and humane society.

Women as strategic self-organizing prefigurative activists

One of the underlying ideas of this volume is that humanity will be faced with difficult choices among an array of wicked dilemmas as we transition to a new kind of low-growth society. Many cherished institutions and ideologies that have arisen alongside growth economics and modernity may come into question as the system transitions into a new phase.

This reminds us that socially just politics and communities have an obligation to "remember." Remembering previous cycles is a part of the panarchy cycle – systems learn and by doing so are less likely to make similar mistakes. Paul Connerton argues that societies can remember social cues that made the world better such as social justice and empathy (1989). This relates to another principle of resilience – to encourage the continuation of learning in a system to know the full extent to which social progress and socio-ecological relationships. This view is in stark contrast to Quilley's degrowth vision as an illiberal agenda (2012, this volume). However, by remembering and early adoption of prefigurative activist identities that are suitable for a post-capitalist system, there is no reason that a society should revert to older ways of thinking based on lower complexity. Rather than seeing a post-capitalist society as a reversal of society, it can be viewed as an evolution of society in which the lessons we've learned can be remembered and brought forward, particularly through prefigurative politics.

One such ideology is feminism and women's emancipation – in a world where the highly energetic services of childcare might not exist, how much room is there for women to really decide to play other roles in society? What does it mean to be a woman in a localized, anarchic, eco-centric, livelihood context?

The emancipation of the modern woman relies on a series of high-energy subsystems that allowed feminism to emerge, such as the birth control pill, state-funded childcare, education systems, and the removal of sexist barriers to work. Through these systems, feminism became inherently wedded with growth and increasing complexity. The relationship between feminism and growth was only necessary after a systemic movement to suppress and control women in society. This suggests that the role of women as powerful and emancipated agents does not necessarily rely on high-energy subsystems. Only women within the context of a world with strong patriarchal roots require such systems for emancipation. By reclaiming the power of the home, rather than getting meaning from the capitalist labour market, one can both combat unsustainable ecological institutions while redefining where power in a new anarchic system comes from. The home is a central ground for action, motivation and life – women should be proud to take ownership of it. A participant in field research regarding changing local economies conducted in Prince Edward Island, Canada, said:

> the most radical thing I've done is quit my job and become a stay-at-home mom. I removed myself from the competition, from seeking self-esteem from my boss, and freed myself from the need for money. And I was judged very harshly for that decision by my feminist friends.

One consequence of modernity is that individualism, cultural relativism, and secularization (Beck, 1992; Giddens, 1973; Grosby, 2013) as explored in Quilley's second chapter of this volume, have undermined cohesive, culturally sanctioned and shared "hero projects" (Giddens, 1990; Kish, Hawreliak, and Quilley, 2016). Identities and methods for developing self-esteem that were once clear have been made murky through the process of modernity, especially as consumption developed into the most common route for the development of self-esteem. For example, being a good mother is no longer simply about ensuring your child is loved and fed. Instead, mothers are bombarded with a host of expert information to help make the 'best' decisions regarding every decision they make and are constantly questioning their abilities – on top of growing cultural pressure to ensure every decision is the best for a child's mental and social development with the least impact on the environment. This has led to higher instances of mothers feeling depressed, uncertain, and confused (Bailey, 1999). In a way, women are also strongly trapped in a rigidity trap. Modern Western women have an immense responsibility to continue to succeed in their traditional domestic roles – in many ways these roles have become more difficult in the last 20 years, as mothering has become increasingly more involved. And while these domestic roles have become increasingly time and mentally consuming, women are also under immense pressure to demonstrate their emancipation through portrayal of a curated individual identity, succeeding in the workforce, and being highly educated.

Modern Western women have an opportunity to develop a more radical understanding of self-ownership; finding the freedom they already have. Women have started to adhere to a societal ideal of what their freedom looks like. Instead of becoming freer, it's restricted a woman's decision to be content in life. If it were more socially acceptable for women to affirm themselves and their own indifference, there might be more women happily choosing one major life project over another – mothering over working, or working over mothering. In the case of Prince Edward Island, women were tapping into this ownness and choosing mothering over working with great satisfaction (Kish, 2018).

Modern Western women are restricted by the limitations prescribed to them in early capitalism, and further weighed down by the supposed emancipation from those limitations. There are many in the world without these limitations on freedom. For instance, in the indigenous community of Haida Gwaii, women are the matriarchs of their community. Even in indigenous communities that are not matriarchal, indigenous women may be "freer" than the modern Western woman, as their roles are both clear and respected. Karen Lewish, an indigenous woman in Southern Ontario, explained that, as a woman, her role in society is very clear to her, but the prescriptive role is not restrictive because a), as it has cultural roots, it's fulfilling as an obligation, and b), if one doesn't want to adhere to the prescribed role, there is a community to support them in finding what they feel they should be doing and what would be fulfilling to them. There is freedom in both the liberal sense of not having to adhere to a role, but a deeper freedom of contentment in the role that is prescribed.

Generally, modern Western women don't have such clear and understood social norms, nor do they have a strong community to encourage exploration of one's need. Instead there is a broad cultural definition of a free, emancipated woman. One possible way out of this, for those that may want to choose mothering over a career, is to develop a radical prefigurative politic of the home – which I call a radical polis-oikos, where women radicalize the home as a ground for eco-feminist action. The grounding assumptions of a livelihood feminism would include a radical, life-giving, care-providing, self-sustaining politic.

Defining exactly what this radical polis-oikos looks like exactly is difficult as systems theory tells us it will be different depending on the context. The more prescriptive the definition of such a movement, the less useful it becomes. However, it might include an adoption of stronger family and community units, acceptance of lower income and therefore reduced consumption, valuation of community volunteerism, establishment of intergenerational care for both elderly parents and younger children, cultivating place-bound identities through relationships, music, arts, and learning, homeschooling, and adoption of more minimalist parenting practices. This would be a transformation in the ideals of feminism – women adopting this radical politic in Prince Edward Island were judged harshly by other women for leaving their high-powered jobs to become full-time, stay-at-home mothers (Kish, 2018).

If more widely adopted, a radical polis-oikos can help begin to carve out a new system where domestic feminism has a strong footing – where women have a strong sense of self, so that in a new system they don't lose their power if the subsystems upholding power (birth control, childcare) no longer exist. Instead, these women can re-occupy the home sphere and be primary definers of what the self-organized system looks like as they'd be the primary decision makers regarding education, food production/making, relationships, and community orientation. This is not to say that all woman should, or need, to do quit their jobs and adopt such a feminism. Instead, those who already do stay at home, are unfulfilled by "bullshit jobs" (Graeber, 2018), or just want a simpler life should be heralded as activists going up against a strong system. The women in Prince Edward Island should feel proud and empowered to choose family over work, not judged, and by establishing a politic around it, they may see the power and importance of their actions. This approach to the rights of women would help broaden modern definitions of an empowered woman that would also contribute to a post-capitalist alternative system.

Moreover – framing motherhood, care, and feminized externalities as grounds for a radical politic provides fodder for policy development around social services, universal basic income, and tax bracketing. Mothering is a strategic and smart grounds for developing an important prefigurative activist agenda to ensure the systemic patriarchy collapses alongside capitalism.

This kind of politic can only really happen in a society that allows for self-organization. However, allowing for self-organization and prefigurative politics will be mutually reinforcing. As communities change to make room for new identities for both women and men, more opportunities to do so would emerge. As more women take on an activist-mother role, society would begin to organize more around the needs of these women – education, food systems, and consumption patterns would change to meet the needs of these new roles. However, this does mean that a great deal of obligation falls onto the shoulders of women to be the instigators of change.

Conclusion

Kropotkin argued that "there are periods in the life of human society when revolution becomes an imperative necessity" (2002, p. 35). We can see this happening when "new ideas germinate everywhere, seeking to force their way into the light" and forcefully "opposed by the inertia of those whose interest it is to maintain the old order" (ibid). In these moments, the need for reorganization and a revitalization of life becomes apparent as the "code of established morality … no longer seems sufficient" (ibid). Thus, the time for reorganization may approach and the strength of the anarchist is their emphasis on self-affirmation, self-organization, and empowerment of the both the individual and the community within which the individual is situated. Self-affirmed individuals in a community where their role matters, may have less of a difficult time with their limited relative freedom, but may be happier than the individual who is free to do whatever they want.

Regardless of orientation and connectedness in the system post-collapse, the system will always find ways to adapt and survive to surprise generators. Catastrophes are regular occurrences as the *system* experiences discontinuity that radically undermines the trajectory of the *system* (such as ecological collapse). The science of complexity challenges us to think differently about how systems move and orient themselves – and how we might begin to define new forms of freedom steeped in self-organization, and learn lessons from what society remembers.

References

Bailey, Donald B., Lynette S. Aytch, Samuel L. Odom, Frank Symons, and Mark Wolery. (1999). "Early intervention as we know it." *Mental Retardation and Developmental Disabilities Research Reviews*, 5(1): 11–20. https://doi.org/10.1002/(SICI)1098-2779(1999)5:1<11::AID-MRDD2>3.0.CO;2-U.

Beck, Ulrich. (1992). *Risk Society: Towards a New Modernity*. London: SAGE.

Berkes, Fikret, Johan Colding, and Carl Folke. (2000). "Rediscovery of traditional ecological knowledge as adaptive management." *Ecological Applications*, 10(5): 1251–1262. https://doi.org/10.1890/1051-0761(2000)010[1251:ROTEKA]2.0.CO;2.

Boggs, David L. (1977). "Visitor learning at the Ohio Historical Center." *Curator: The Museum Journal*, 20(3): 205–214. https://doi.org/10.1111/j.2151-6952.1977.tb01638.x.

Breines, W. (1980). "Community and organization: The new left and Michels' 'Iron Law.'" *Social Problems*, 27(4): 419–429. doi:10.2307/800170.

Connerton, P. (1989). *How Societies Remember*. Cambridge, England, New York: Cambridge University Press.

Cornish, Amelia, David Raubenheimer, and Paul McGreevy. (2016). "What we know about the public's level of concern for farm animal welfare in food production in developed countries." *Animals: An Open Access Journal from MDPI*, 6(11). https://doi.org/10.3390/ani6110074.

Epstein, Joyce L., and Susan L. Dauber. (1991). "School programs and teacher practices of parent involvement in inner-city elementary and middle schools." *The Elementary School Journal*, 91(3): 289–305.

Erlic, M. (2016, November 26). "What is the adjacent possible?" *Medium.com*. Retrieved August 18, 2019 from https://medium.com/@SeloSlav/what-is-the-adjacent-possible-17680e4d1198

Giddens, Anthony. (1973). *Capitalism and Modern Social Theory: An Analysis of the Writings of Marx, Durkheim and Max Weber*. Cambridge: Cambridge University Press.

———. (1990). *The Consequences of Modernity*. Stanford, CA: Stanford University Press.

Gotts, N. (2007). "Resilience, panarchy, and world-systems analysis." *Ecology and Society*, 12(1). doi:10.5751/ES-02017-120124.

Graeber, David. (2009). *Direct Action: An Ethnography*. 1st edition. Edinburgh: AK Press.

———. (2018). *Bullshit Jobs: A Theory*. New York: Simon & Schuster.

Grosby, Steven. (2013). "Max Weber, religion, and the disenchantment of the world." *Society*, 50(3): 301–310. https://doi.org/10.1007/s12115-013-9664-y.

Gunderson, Lance H., and C. S. Holling, eds. (2001). *Panarchy: Understanding Transformations in Human and Natural Systems*. 1st edition. Washington, DC: Island Press.

———. (2002). *Panarchy: Understanding Transformations in Human and Natural Systems*. Washington, DC: Island Press.

Holling, C. S. (1973). "Resilience and stability of ecological systems." *Annual Review of Ecology and Systematics*, 4(January): 1–23.

Holloway, J. (2010). *Crack Capitalism*. London: Pluto Press.

Ingold, T. (2011). *Being Alive: Essays on Movement, Knowledge and Description.* London, New York: Routledge.

Janssen, Marco, Örjan Bodin, John Anderies, Thomas Elmqvist, Henrik Ernstson, Ryan R. J. McAllister, Per Olsson, and Paul Ryan. (2006). "Toward a network perspective of the study of resilience in social-ecological systems." *Ecology and Society, 11*(1). https://doi.org/10.5751/ES-01462-110115.

Kish, K, S Quilley, and J Hawreliak. (2016). "Finding an alternative route: Towards open, eco-cyclical and distributed production." *Peer Production, 9.* http://peerproduction.net/issues/issue-9-alternative-internets/peer-reviewed-papers/finding-an-alternate-route-towards-open-eco-cyclical-and-distributed-production/.

Kish, K. (2018). "Ecological economic development goals: Bringing the social sphere back into ecological economic imagination." *UW Space*, Dissertations, 1.

Kropotkin, P. (2002). *Anarchism: A Collection of Revolutionary Writings.* edition. Mineola, New York: Dover Publications.

Lockyer, Joshua, and James R. Veteto. (2013). *Environmental Anthropology Engaging Ecotopia: Bioregionalism, Permaculture, and Ecovillages.* New York: Berghahn Books.

Maldonado, C. (2016). "Anarchy and complexity." *Emergence: Complexity and Organization, 18*(1): 52–73.

Marx, K. (1844). *The Economic and Philosophic Manuscripts of 1844 and the Communist Manifesto.* 1st edition. Amherst, NY: Prometheus Books.

Newman, S. (2016). "Whither anarchy: Ownness as a form of freedom." *The Conversation.* Retrieved from http://theconversation.com/whither-anarchy-ownness-as-a-form-of-freedom-60777.

Polletta, Francesca. (1999). "'Free spaces' in collective action." *Theory and Society, 28*(1): 1–38. https://doi.org/10.1023/A:1006941408302.

Porta, Donatella della, and Mario Diani. (2006). *Social Movements: An Introduction.* 2nd edition. Malden, MA: Wiley-Blackwell.

Rockstrom, Johan, Will Steffen, Kevin Noone, Asa Persson, F. Stuart III Chapin, Eric Lambin, Timothy M. Lenton, et al. (2009). "Planetary boundaries: Exploring the safe operating space for humanity." *Ecology and Society, 14*(2): 32.

Schumpeter, J. A. (1942). *Capitalism, Socialism and Democracy.* 1 edition. London, New York: Routledge.

Schumpeter, J. A. (1947). "The creative response in economic history." *The Journal of Economic History, 7*(2): 149–159. doi:10.1017/S0022050700054279.

Smith, E. (2010). *The Adjacent Possible.* Practically Efficient blog. www.edge.org/conversation/stuart_a_kauffman-the-adjacent-possible

Tainter, Joseph A. (1988). *The Collapse of Complex Societies.* Cambridge, UK; New York: Cambridge University Press.

Walker, B. H., & Salt, D. (2006). *Resilience Thinking Sustaining Ecosystems and People in a Changing World.* Washington, DC: Island Press.

Westley, F., Zimmerman, B., & Patton, M. (2007). *Getting to Maybe: How the World Is Changed.* Reprint edition. Toronto: Vintage Canada.

Williams, S. J. (2017). "Personal prefigurative politics: Cooking up an ideal society in the woman's temperance and woman's suffrage movements, 1870–1920." *The Sociological Quarterly, 58*(1): 72–90. doi:10.1080/00380253.2016.1246894.

Young, K., & Schwartz, M. (2012). "Can prefigurative politics prevail? The implications for movement strategy in John Holloway's crack capitalism." *Journal of Classical Sociology, 12*(2): 220–239. doi:10.1177/1468795X12443533.

Seeds of freedom and nature in modern traditions

PART II

Seeds of freedom and failure in modern traditions

5

ARE FREEDOM AND INTERDEPENDENCY COMPATIBLE? LESSONS FROM CLASSICAL LIBERAL AND CONTEMPORARY FEMINIST THEORY

Amy R. McCready

A dominant understanding of freedom is the absence of impediments to the actions of self-interested individuals conceived in isolation from one another. People are free when they can pursue their preferences without interference from "outside" entities, and they should be free because they are rational autonomous agents who can choose their own ends. This conception of freedom, called "negative liberty," and the ontology on which it relies are arguably components of liberal theory, and have been criticized as such.[1] They are also associated with the cultural ideal of self-sufficiency and the "rational actor" of capitalism, ideas that have contributed to the Anthropocene.

The valorization of self-sufficiency, crystalized in the notion of the "self-made man" in the West, casts freedom as the antithesis of dependency. To be free, one cannot be dependent upon or indebted to others, so those who are do not qualify as free and are hence denied equal political participation and/or social recognition.[2] Because dependency is denigrated and defined in opposition to freedom, it is feared and disavowed (Fineman, 2004; Tronto, 1993). This dichotomy between freedom and dependency became gendered in modernity as commercial economies severed "work" from the household and equated wages with value, making women economically dependent and their labour in the home insignificant if not invisible. Men's dependency on women's work in the private sphere, as well as elites' dependence on lower classes and humans' dependence on the natural world, are conceptually incompatible with this version of freedom and thus concealed.

This "free" individual's inability to acknowledge both his reliance on others, and the possibility that "dependents" have ends of their own, is extended in advanced capitalism. Once reason is reduced to rational calculation, social interactions are regarded as self-interested exchanges, and antagonism between market actors is assumed, even others who are recognized as equal and free – the elite's peers – are treated instrumentally, as means to one's own ends, rather than as agents accorded intrinsic value. In contrast to negative liberty, which defines restraints on freedom as an external coercive force, and liberal theory, founded on justice, capitalism dispenses with ethical obligations as undue impediments to freedom. Further, self-interest in this system threatens equality because freedom may entail domination (its opposite in other accounts), which the competitive market sanctions. Capitalism's demand for continual growth goads consumption, as does "success" measured by material goods, such that this version of freedom that idealizes individualism and self-interest while rejecting dependency is at odds with environmental sustainability.

Given our dependence on nature and our ecological crisis, a crucial conceptual task is to identify whether, and if so how, freedom and interdependency may coincide.[3] What conceptions of freedom encompass dependencies and obligations to persons and the environment, versus that which opposes freedom to dependency and ethical relations with others? This chapter offers options by examining the ontologies and understandings of freedom in the 17th-century political theory of John Locke and works by contemporary feminist theorists. More specifically, the chapter first employs two ideas emphasized by these feminists, namely, that dependency is a fundamental quality of life neglected in mainstream views of freedom, and that knowledge and ethics are contextual, to probe Locke's thought. If dependency is empirically undeniable, so that the idea of free individuals fosters "a false picture of society" (Held, 2006, p.14), how does Locke seemingly interpret it away, and, most importantly, why? To answer these questions, the chapter analyzes Locke's political theory in the context of monarchical rule, justified by patriarchal theory, which he sought to refute and replace with representative government. Although Locke casts his theory as universal, this chapter treats it as historically specific.

The second half of the chapter explains strands of feminist theory – the ethics of care, ecofeminism, and a social constructionist conception of liberty – that directly advocate a relational ontology in place of the autonomous agent of modern liberalism and neo-classical economics. Although their conceptions of the relational self vary, all emphasize our reliance on others, the constitutive quality of contexts, ethical obligations, and a positive holistic view of many human qualities, versus the denigration of dependency, ahistorical assumptions, the instrumentalization of others, and a fixation with a narrow view of rationality characteristic of the conception of freedom outlined above. The critical question that this chapter poses to these feminists is whether a relational ontology is necessarily in tension with freedom. What formulations of freedom may be compatible with dependency/interdependency and the cultural constitution of identity and reality? The chapter analyzes the conceptions of freedom that are criticized and endorsed (whether directly or in other terms) in these three variants of feminism to show that, in spite of their different aims, each recognizes the need for independence within interdependent lives. Having explained a range of relationships between dependency and freedom in

Locke and recent feminist research, the chapter draws upon the ethical insights of the latter to highlight obligations to others and the environment, which are both conceptually entailed by a relational ontology and imperative given ecological peril.

From natural subjection to Lockean freedoms

Before going back centuries, why does Locke deserve this attention? First, the conception of freedom described above is thought by some to derive from early modern liberal theory, particularly that of Locke. This attribution is not absurd, for Locke defended the natural freedom of individuals based upon reason, so that legitimate relationships required consent and took a contractual form, and declared nature worthless until developed by labour in his justification of private property. It is, however, historically simplistic and exegetically unfounded. Second, Locke articulates an ethical individualism, moral norm of our "proper Interest," and interdependency among persons – all essential to *"perfect Freedom"* (Locke, 1960, II.57, II.4) – which are themselves alternatives to freedom as egoists' unlimited pursuit of self-interest in capitalism.[4] Third, Locke simultaneously debases nature and acknowledges our absolute reliance on it. The ecological import of his ideas is skewed if we use our culture's connotations of freedom and dependency instead of his.

As noted above, the cultural and conceptual framework of classic patriarchy, with which Locke is explicitly in critical dialogue, is crucial for this analysis, so understanding its contours is essential.[5] Patriarchal society was justified with reference to the hierarchically arranged cosmos, a chain of beings descending from angels to humans to fauna and flora. The divine and nature converge in patriarchal theory because God created the world in such differentiated parts. The inequalities patently evident in nature, representing God's unassailable will, verified the superior–inferior relationship of ruler and subjects. King James (1994) cleverly used the hierarchy of head over body to mock rebellion, but the natural inequalities in the family were the central organic foundation of patriarchal government.[6] As humans are undeniably born helpless, in a state of subjection to paternal power, so subjects are dependent upon the king. As fathers in fact wield the power of life and death over their dependents, so the power of the monarch is absolute.[7] While we associate such power with outright oppression, it was ethically founded in this world-view and imposed responsibilities. The superior was obliged to provide for and protect the inferior, who correspondingly was obliged to obey.[8] Reproduction grounded monarchical power not only because the ruler–subject relationship replicated that of father and child, but also because the right to the throne was transmitted by blood. Given that God our omnipotent father granted particular men the power to rule, they literally embodied it, so their lineage was sanctified. The irrefutable empirical evidence of nature – sexual intercourse breeding entirely dependent young – justified monarchy's ethic of essential inequality.

Propounding an alternative, Locke construes the relationship between freedom and dependency in several ways, which correspond to the state of nature, private property, and conjugal relations (marriage). All three serve the preservation of the species, but vary in aspects relevant to his dichotomy between humans and nature. The first articulates a conception of freedom as agency, the power to act by will or choice, which entails interdependency in this chapter's analysis. The second deploys negative liberty – freedom from interference by others – when reliance on them (dependency) could threaten survival. The third combines appetite with agency to ensure that dependent infants by nature become free. The connections between freedom and dependency – and their meanings and relevance – are discussed in this order.

Given patriarchy's grounding in natural inequalities, Locke redefines nature. The foundation of his theory is the state of nature, "*a State of perfect Freedom*" in which persons live "as they think fit, within the bounds of the Law of Nature [reason], **without** asking leave, or **depending** upon the Will of any other Man" (Locke, 1960, II.4, bold added). His description of equality likewise counters monarchy: it is an "*equal Right*" of each "*to his Natural Freedom*, **without being subjected** to the Will or Authority of any other Man" (Locke, 1960, II.54, bold added). Locke's linked conceptions of freedom and equality are forged in opposition to patriarchy's interpretation of dependency as subjection. Natural equality also means that "all the Power and Jurisdiction is reciprocal, no one having more than another" (Locke, 1960, II.4). The power that absolutists asserted – over life and death – belongs only to God, thus prohibiting subordination, a form of inequality that instrumentalizes others (treats them "as if we were made for one anothers uses" [Locke, 1960, II.6]). Liberated from organically conceived hierarchies, humans' capacity to choose comes to define the species and justify rights. But freedom is contingent upon a social foundation, shared knowledge of the law of nature. Unless most consult reason, our "Star and compass" (Locke, 1960, I.58), we cannot experience the freedom for self-government or choose the particulars of one's life. Freedom is necessarily a relational good.[9]

When dependency is construed as inferiority and subjection, it is antithetical to freedom, but Locke's conception of freedom relies on dependency on moral equals. His explanation of self-preservation also evinces interdependency, further distinguishing his theory from the view of freedom as self-sufficiency that feminists rightly malign. "Every one as he is *bound to preserve himself* ... so by the like reason when his own Preservation comes not into competition, ought he, as much as he can, *to preserve the rest of Mankind*" (Locke, 1960, II.6). The inclination of each to ensure his well-being intrinsically entails an obligation to preserve others, and to do so thoroughly, "as much as he can." Self-interest and that of the species coincide. Individuals in Locke's state of nature are not only not adversaries, but, moreover, act for the benefit of humanity; they understand that their lives are inherently intertwined. Reason directs "*free and intelligent Agent*[s]" to their "**proper Interest**," that which serves "**the general Good**," our collective existence (Locke, 1960, II.57, bold added).

Locke identifies freedom with seemingly exclusively human abilities associated with the mind – intelligence and reason – so we are ethically bound to our species alone, helping to obliterate obligations to the environment for centuries. He does so, however, to liberate us from the embodied inequalities and subjection to natural superiors that patriarchy rooted in organic hierarchies. Because nature was ethically determinative in patriarchal rule – justifying superior/subject relations – Locke deprives it of ethical significance. He severs the links in nature's chain of beings that connected humans, animates, and vegetative life to argue for freedom from absolutism and for self-government.

Although Locke eliminates the inequalities between persons that pervaded patriarchal societies, he transposes superior/inferior to humans versus others, a distinction of devastating importance to the Anthropocene. His demotion of nature from ethical foundation to matter without value is clearest in his explanation of property. "Land that is left wholly to Nature, that hath no improvement of Pasturage, Tillage, or Planting, is called, as indeed it is, *wast*; and we shall find the benefit of it amount to little more than nothing" (Locke, 1960, II.42). Estimating in useful goods "what in them is purely owing to *Nature*, and what to *labour*, we shall find, that in most of them 99/100 are wholly to be put on the account of *labour*" (Locke, 1960, II.40), hence environmentalists' condemnation of Locke.

The labourer, however, is not freely acting, for "his Wants forced him to *labour*" (Locke, 1960, II.35), "the penury of his Condition required it of him" (Locke, 1960, II.32). The exclusively human capabilities essential to freedom – reason and choice – seem irrelevant to procuring material resources, while our embodiment and animality, absent in Locke's conception of freedom, are central. Yet like freedom, his account of property hinges on refuting patriarchs' power over subjects' life and death. Given that labour is exerted to survive physically, unilateral acquisition is justified, for the alternative, depending for nourishment on anyone's will, would put self-preservation at risk. If "the consent of all Mankind" or "of any body" were required prior to appropriation, then "Man had starved" (Locke, 1960, II. 28). Consent in this case seems tantamount to subordination and crime, destroying the "Property" and "Workmanship" of God (Locke, 1960, II.6). Because we are vitally dependent upon the earth, access to it cannot depend on other people.

Even as Locke asserts nature's insignificance, its meagre value relative to ours, he emphasizes our ultimate reliance upon it. Far from denying our dependency on nature, he justifies private property via this fact. Paradoxically, an idea that has contributed to capitalist exploitation of the environment is based in Locke's recognition that we need it to stay alive. Explaining property, Locke derogates nature, but does not unequivocally elevate humans above other sorts. Our vulnerability – our embodiment – drives his defence of individual appropriation of material resources, and markedly distinguishes it from his conception of freedom which is intellectually founded and necessarily relational as explained above. When freedom to live by reason is at issue, dependence on peers is essential, but when survival is at stake, being dependent on another could mean death. Although liberal states and capitalist economies are historically linked and thought to be undergirded by the same conception of freedom, political life and private property are justified very differently in Locke's account. Freedom in the former is a positive relational human good, while the freedom pertinent to property is freedom from others' possible interference when we work to stay alive. Labourers are not agents, able to act or not by their own lights. Our animality and dependence on nature, not freedom, are ontologically primary in Locke's justification of private property.

In contrast, freedom seems foremost in his account of marriage, for "*Conjugal Society* is made by a voluntary Compact" (Locke, 1960, II.78), but his dialogue with patriarchal thought reveals a more complicated view.[10] Because patriarchies conceived of men as naturally superior to women, the logic of Locke's argument against natural inequalities extended to sex along with status. Because humans are naturally free and equal, legitimate relationships required consent. While conjugal society thus seems to reflect agency, the appetites operative in reproduction in other species are actually decisive in Locke's account, challenging patriarchy's equation of human reproduction with will and sovereignty. When copulating, humans behave as beasts do, not only without but often against the participants' consent.

> What Father of a Thousand, when he begets a Child, thinks farther then the satisfying his present Appetite? God ... has put strong desires of Copulation into the Constitution of Men ... to continue the race ... which he doth most commonly without the intention, and often against the Consent and Will of the Begetter.
>
> *(Locke, 1960, I.54)*

The freedom suggested by the contract is not agency – whether to have sex or not – but only with whom, and its content underscores the corporeal core of the bond in Locke's view: the agreement consists "chiefly in such a Communion and Right in one anothers Bodies, as is necessary to its chief End, Procreation" (Locke, 1960, II.78).

Moreover, Locke compares human parents with those of other species to set the duration of the relationship, reinforcing our animality rather than raising us above other breeds. As "Cock and Hen continue mates, till the Young are able to use their wing, and provide for themselves" (Locke, 1960, II.78), so humans remain together until their progeny mature, after which they may part. Situating us among other animates also enables Locke to counter patriarchy's empirical riposte to his idea of natural freedom, namely, our subjection at birth to natural superiors. The offspring of most animate species depend on parents, but they also thereby naturally develop and depart. All creatures are by nature set free. While natural freedom seems to be the antithesis of natural subjection, it is actually, as evident across species, its natural result. When children are viewed like cubs and fledglings, we see how natural processes turn them into adults. The relationship between childhood dependency and freedom is a matter of tutelage and time.[11]

Thus far, this chapter has analyzed Locke's political theory via the patriarchal context in which it was forged to identify the ways in which dependency and embodiment, qualities of the human condition that contemporary feminist theorists emphasize, are implicated in his conceptions of freedom. Because the ostensibly free agents in his understandings of natural society, property, and marriage are either interdependent or compelled to act by physical needs, they differ from the notion of self-sufficient rational actors in capitalist culture. These elements of relationality and vulnerability do not reconcile liberal and relational ontologies or obviate feminist challenges to individualism, but they yield a fuller understanding of the meanings of freedom and the centrality of dependency in this classic in Western political thought.

Relational freedom and responsibility

Dependency within relationships is central to many feminists' analyses not only because it is inescapable, but also because it is disparaged and obscured as noted above. Fineman (2004, Chapter 2) examines the "inevitable dependency" of children, the ill, and elderly, which entails the "derivative dependency" of caretakers who need material resources but cannot fully participate in the labour market, making them vulnerable to "free" (male) breadwinners. By assigning caretaking to the private family, liberal-capitalist states disclaim responsibility for the costs of this "society preserving" work, incurring, Fineman argues, an unacknowledged collective debt. Held (2006) explains that even liberals must recognize the "deeper reality of human interdependency" (p. 43) on which their free individuals and contractual relationships depend. Liberal selves – reasonable adults – must be developed from children; citizens are formed through the particular ties and values of caring homes.[12] Held also incisively identifies the flaw in rights and contracts as legal norms for interactions between strangers. Prior to the social contract (and any contract), persons must agree upon with whom they will agree. "All must feel sufficiently connected to seek agreement among themselves to respect each others' rights" (Held, 2006, p. 129). Contract presupposes community. Freedom is the product, not the precursor, of a shared social life. Lastly, ecofeminist Plumwood (1993) identifies the "denial of dependency" on "the whole sphere of reproduction and subsistence" as "a major factor in the perpetuation of the non-sustainable modes of using nature which loom as such a threat to the future of western society" (p. 21). It generates conceptual oppositions between reason and nature, mind and body, freedom and necessity, which must be dismantled to reveal the continuities between nature and the human, and their integral interdependence, requisite for ecological change.

Contemporary feminist scholars also stress the epistemological significance of cultural contexts, criticizing approaches in ethics that abstract from particularity (of place, experience, relationships) to produce ostensibly universal norms.[13]

They expose the male-bias in the dominant view of freedom, explaining how its claims of objectivity include patriarchal assumptions, enable the powerful to secure their partial perspectives under the guise of value-neutrality, and thus perpetuate patriarchal systems. In contrast, feminists contend that "facts" include normative judgments; "reality" is produced through discursive frameworks, formal and informal social structures, and iterative interactions. Care ethicists like Tronto (1993), ecofeminists such as Merchant (1980), and feminists who study freedom like Hirschmann (2003) deeply attend to the ways in which specific historical contexts yield particular values, variants of self and other, and visions of society, and, correspondingly, conceal others.

Like Kish and Quilley (both in this volume), these scholars challenge liberal feminism's adoption of individual freedom and formal equality as routes to inclusion, and advance relational ontologies. But the ethic of care, ecofeminism, and feminist freedom examined in this chapter base the selves and values that they endorse in women's experiences. The idea that identities are formed within networks of relationships (e.g., families, communities, friends, regions, citizens) is significant for several related reasons. This summary, numbered only for clarity, contrasts feminist and capitalist ontologies. First, individualism and freedom may be generated by social networks, but are not antecedent to them. Interdependency is fundamental. Secondly, ties and interactions with others constitute who we are, such that values and identities are not primarily matters of individual choice. Relationships are not optional, inessential, and temporary as contract models suggest, but core sources of our "inner" selves and actions. Third, because our lives are shared with others, our interests are intertwined, not intelligibly separable or at odds as contractual exchanges assume. One's well-being is affected by the experiences of his/her intimates, communities, and environment. Fourth, embedded as we always are in relationships, ranging in scale from homes through global societies, we are obliged to others from the start. Our responsibilities are not limited to those to which we consent, as the notion that relationships are contracts between atomistic individuals asserts. Unchosen ties with parents, siblings, and other relatives, which endure for a lifetime for most, convey obligations. Our vital connection to nature similarly entails unchosen responsibilities to the planet. Finally, feminist theorists endorse a holistic ontology, valuing affect and embodiment not only as qualities of life neglected in liberal-capitalism's focus on rationality, but also as sources of moral judgment and action without which reasoned argument may be sterile.

The values advocated by both the ethic of care and ecofeminism follow this ontology.[14] Given the significance of relationships, maintaining them through cooperation, trust, dialogue, compromise, and respect for differences (versus mere toleration) is paramount, as evident in friendships. In contrast to adversarial interactions between solely self-interested individuals, the relationship itself is positively valued and hence preserved. A key part of doing so is attending to the particular needs of others who are dependent upon us (and on whom our well-being relies) and acting to alleviate their suffering and support their flourishing.

The qualities of effective caring are valued in turn: sensitivity, sympathy, responsiveness, judgment, and sincere concern. These contrast with the values associated with contracts: rational calculation, freedom for one's own pursuits, and competition as a route to progress.

While care ethicists draw upon the experiences and moral perspicacity of mothers caring for children, they argue that care should be our political priority. Held (2006) elaborates ways in which "it is a radical ethic calling for a profound restructuring of society" (p. 19), but does not include the natural world as ecofeminism does. As mentioned above, a conception of freedom that denies and disparages dependency is a key culprit in these feminist critiques. The ontological assumptions of capitalism, that all persons are free and equal such that self-sufficiency is venerated and dependency shamed, combined with the privatization of the work of care (to homes and private-sector health care provided through employment in the United States), foster moral indifference to the lives of others and society at large (Held, 2006, p. 83). This variant of freedom absolves one from any sense of social responsibility, for obligations are confined to those explicitly chosen. A privileged irresponsibility prevails, certainly among the elite (Tronto, 1993), but more widely if we think that responsibility for the environment should be incumbent on all.

Correspondingly, care ethicists argue that looking beyond what we usually consider to be "our" concerns is morally imperative. Paying attention to others' needs and the conditions of our shared existence, and recognizing responsibility for them, must be moral expectations. Focusing exclusively on one's personal well-being is consequently a moral failing. A direct implication of this ethic, though one not drawn by Held, is that neglecting to understand and act upon climate change is an inexcusable vice. As members of the community of earth, we are already obliged, as a matter of morals and not only necessity, to attend to its condition, meet its needs, and actively care for the whole.

If mutuality and care become our highest social concerns, is freedom displaced from the centre or even discouraged outright? Is relationality necessarily in tension with freedom, or only with its dominant definition? Held (2006) usually reserves the term for the "abstractly independent, free, and equal individuals" conjured by liberalism as "self-sufficient" (p. 84, 86). It is freedom from others that "promotes only calculated self-interest and moral indifference in place of the caring and concern that citizens often have for fellow citizens" (Held, 2006, p. 83). Held (2006), however, proposes retaining "justice," which "protects equality and freedom" through individual rights, in part to expand "the positive rights of persons to what they need to act freely" (p. 15, 69). Rights can be a vehicle for care if "the traditional liberal view that freedom is negative only" (Held, 2006, p. 69) is dispatched and social cooperation is accepted as empirically and morally primary.

Held (2006) also thinks that care should develop persons' capacity for "autonomous choices," stressing that autonomy "is exercised within social relations" (p. 84), not by abstract free agents. Given that autonomy is often thought to require deliberation among distinct options and reasoned defence of one's preferences, it seems laden with the rationalist and individualist tendencies that she rejects. Her purpose, however, is to acknowledge the importance of space within social ties. The "autonomy sought within the ethics of care is a capacity to reshape and cultivate new relations" (Held, 2006, p. 14), representing a need for persons to be free, one might say, to evaluate and form communities for themselves.

In contrast to Held, Plumwood (1993) offers an ecofeminist perspective that seeks "a coherent liberatory theory" (p. 13) for the green movement that encompasses oppression in all forms. Yet freedom is not the desired outcome, for she associates "freedom" with a domination of others and psychic distance from them, via reason defined in opposition to embodiment, nature, and emotion (Plumwood, 1993, pp. 23, 28, 53). Like Held, Plumwood argues that reality has been distorted by an ontology starring free rational egoists who deny their own dependency while declaring that passivity and necessity inherently anchor the existence of others, who are thus not regarded as agents with their own ends. Consequently, the egoist treats them as means to his goals. Through the agent/non-agent distinction, he "erases the other as part of the ethical domain" (Plumwood, 1993, p. 145), precluding respectful and empathetic interactions.

Plumwood (1993) hence aims to liberate us from the conceptual straightjacket of dualism, which is "the process by which contrasting concepts" such as mind/body or culture/nature "are formed by domination and subordination and constructed as oppositional and exclusive" (p. 31). Dualism polarizes "identities so as to make equality and mutuality literally unthinkable" (Plumwood, 1993, p. 47). She disrupts the hierarchical dualist dynamic, which supports intra-human oppression and our domination of nature, by exposing the continuities between what Western thought has cast as categorical differences. She restores agency and intentionality to other species and ecosystems, as well as valuing our animality, emotion, and imagination, reintegrating qualities that have been unduly divided. If we recognize that all in nature have needs, strive for ends of their own within relations with others, and are "creative, self-directed, and originative" (Plumwood, 1993, p. 124), then we may develop mutuality with them.[15]

Although Plumwood uses the language of kinship, community, solidarity, and friendship, stressing that relationships are essential (not contingent), she realizes that those instrumentalized by egoists need independence too. Having analyzed five means by which master-other relations construct one as agent and the other as inferior – backgrounding, hyperseparation, incorporation, objectification, and homogenization – she outlines a solution for a "relationship of non-hierarchical difference" (Plumwood, 1993, p. 60) To remove the first two of these problems, "agents" must acknowledge their dependency on and continuity with others, and to alter the remaining three, those dominated must be freed. We must reclaim their "positive independent sources of identity and affirm resistance," recognize that their "ends and needs are independent ... and to be respected," and grant their "diversity" (Plumwood, 1993, p. 60). Like Held, Plumwood values autonomy and agency as long as they are conceived within relationships instead of being defined against them. While continuity among life forms and our interdependent interests should generate collective care, independence and individuation allow each to exercise a critical kick. Although Plumwood (1993) favours "mutual and sustaining interchange with nature" and "active dialogue with earth others" (p. 137, 139), she nonetheless acknowledges that relational respect also requires remove. The "ecological self recognizes the earth other as a centre of agency or intentionality having its origin and place like mine in the community of the earth, but as a different centre of agency, which limits mine" (Plumwood, 1993, p. 159). Agency allows meaningful engagement, but also insists upon space. Being free from the interference of others and free to flourish in one's own way are components of the relational value of care, even though Plumwood refrains from using "free."

In contrast to the ethics of care and ecofeminism, Hirschmann offers a full-throated defence of freedom and choice – long denied to women – within feminist theory. Like Held and Plumwood, she emphasizes the social construction of identities and values, and charges the dominant strain of freedom with misrepresenting reality, producing gendered hierarchies covertly under the banner of equal rights, and setting the conceptual parameters through which we think. Negative liberty assumes that persons are impervious to "external" influences like cultural norms or relationships, so it recognizes only overtly coercive forces (law or physical power) wielded by discrete agents as limits to freedom. Aside from such restraints, individuals are free, and their choices represent their rational will and interests. Given this atomistic ontology, contracts create connections with others and confine accountability to explicit agreements under one's control (in contrast to the pre-existing ties that convey obligations for relational selves).

Hirschmann counters this abstract individual with a self "internally" constituted through particular "external" frameworks, such that desires and choices reflect cultural ideologies and the institutions that they allow. "We must acknowledge the *interaction* of 'inner' and 'outer' and see them as interdependent in meaning and in practice" (Hirschmann, 2003, p. 13). Freedom may thus be limited by both "internal" barriers (e.g., socialization) and structures such as patriarchy and capitalism that are immensely powerful yet not identifiable as intentional agents and hence not obstacles to freedom in the orthodox view.

Although these social forces establish inequalities, Hirschmann stresses that we all regenerate power relations, often unconsciously, through assumptions and ordinary interactions. Some have more power than others, but all participate cognitively, affectively, and behaviourally in replicating social systems; these exceed control by any group. More contentiously, she argues that those who are most vulnerable to oppression, whose options are severely constrained, must be free to choose and should have their choices respected, even if they are "irrational" in another's view. Emphasizing that freedom is contextual, relative, and contingent, Hirschmann defends choice even when patriarchy seems to be in full force. To do otherwise, she argues, risks paternalistic power and vacates the portion of freedom that ought to be preserved.

Although Hirschmann thus validates a version of individualism – persons should have authority over themselves – she pairs freedom with responsibilities to others, which the dominant connotations of these concepts evade.

> [N]egative liberty defines itself in opposition to concepts such as obligation and authority; these things, while perhaps necessary to human society, or even to individuals' pursuit of their desires and possibly even to greater freedom in the future, are nonetheless limitations on freedom.
>
> *(Hirschmann, 2003, p. 5)*

Her solution is not simply to reattach responsibility to choice (without which it seems morally unintelligible), but to extend it to relationships. She argues that a battered woman must be free to remain with her abuser, but the dynamic between patriarchal conditions and "internal" desires obliges others to question her, linking relationality with freedom:

> I have an obligation to continue asking, for being exposed to the questioning – being 'subjected' to it, if you will – is necessary to freedom, or at least to being as sure as possible that the conditions for personal choice have been maximized, if not attained.
>
> *(Hirschmann, 2003, p. 236)*

This "dialogic critical questioning," made effective through diverse views, is our responsibility; "[t]hough freedom must be expressed by individuals, its conditions are made possible by community" (Hirschmann, 2003, p. 237).

As a social constructionist, Hirschmann rejects individualism when identifying barriers to freedom not only at a personal level – others' questions challenge the confines of a choice-maker's view to facilitate her freedom – but also societally, for "personal" problems are precipitated by oppressive structures. Accordingly, she expands responsibility with regard to those who qualify as perpetrators and victims in the dominant view. As noted above, negative liberty assigns blame to discrete purposeful actors who have causally violated another's freedom. Hirschmann convincingly argues that freedom is systematically barred by discourses such as patriarchy, capitalism, and racism, so they are "agents" responsible for our plight at which we should take aim. Recognizing that cultural systems are impediments to our ends and that we are unintentionally complicit in them, we are responsible for piercing their power, for working collectively and persistently to generate new meanings, practices, and contexts.

In sum, this section shows that even one like Plumwood, who argues that liberal freedom subverts itself by devolving into the subordination of others, acknowledges the need for independence within relationships. And those like Hirschmann, most committed to the value of freedom, especially for those who have little experience of it, decry the disavowal of responsibility in the dominant conception of freedom and demonstrate the ways in which that conception vitiates freedom substantively. All three theories discussed here condemn the idea of isolated individuals bound to others only contingently through voluntary agreements, a powerful illusion that undergirds capitalism and has distorted our perception of reality. They also, albeit in different ways, argue that freedom is morally required within relationships and essentially includes responsibilities to others, consistent with Locke. Yet they defend much more robust understandings of responsibility that promote values such as care and awareness of social ills, and significantly extend our obligations to others and our accountability for the conditions in which we live. Their relational alternatives to the status quo thus provide a foundation for ecological consciousness and action.

Conclusion

In Locke's time, patriarchy's conceptions of natural hierarchies and dependency as subjection were key conceptual and political problems. In response, Locke articulated a conception of freedom grounded upon a social foundation, standard of "proper Interest" as "the general Good," and mandate to preserve not only self but the species.[16] These linked ideas are crucial to retrieve not only because of their provenance but because they force us to ask whether "freedom" without equality and ethics, without relationality and responsibility, is intelligible or viable. In our time, the variant of individuation that breeds the haughty delusion of self-sufficiency, a conception of freedom bereft of interdependency and historical context, and contractual models of relationship and obligation have contributed to the Anthropocene. To counter these jointly conceptual and political problems, the feminists' relational ontology, grounded on our inherent ties and intertwined interests with others and the earth, is critically important. Because it is evident in everyday experience, it is often endorsed empirically and ethically when raised to the surface, at least among 21st-century undergraduates. Moreover, it offers an understanding of difference less prone to domination than modernity's ideal of inclusion via abstraction from our situations in bodies, time, and place. This ontology also recognizes the contingency of freedom, calling for attention to the circumstances of "choice," which must now include the carrying capacity of our planet.

Drawing upon these scholars' work, and given the Anthropocene, three interrelated modifications of the ethical individualism at the conceptual foundation of liberal democracies are warranted. First, because we live within many vitally important relationships and communities, and are dependent upon them in myriad respects, we are always and already obliged to others. Responsibilities precede us – they are not limited to those we may opt to incur – and they include learning about others' diverse ends. We have a moral obligation to pay attention to the needs of other beings, including ecosystems, and respond to them with a sensitivity to their contexts and well-being (versus our view of their interests), granting that they may need more freedom. Second, in addition to responding to concerns that demand our attention, we should deliberately broaden our scope, seeking to discern vulnerabilities that should be ameliorated and delete-rious social and ecological conditions that should be redressed. Insularity and indifference are morally irresponsible; conscious attention to – caring about – the worlds "outside" of our "own" is incumbent upon us, as is acting upon what we see. And what confronts us when we look more searchingly are systemic injustices supported by interpretive frameworks that blind us to their power. So, third, acknowledging that our perspectives are constructed through destructive discourses and the social systems that they support, we should take responsibility for our complicity in them, for recognizing them as impediments to substan-tive freedom, and for struggling to reconstruct them in more sustainable forms. Comprehending the dynamics through which damaging cultural practices work their way through us, such that we are implicated in their harms, obliges us to think and act anew to cultivate contexts that we would choose to endorse.

Notes

1 Hirschmann (2003) explains Isaiah Berlin's distinction between negative and positive liberty.
2 See Quilley's chapter for his analysis of the state's role in individuation, liberty, and market economies.
3 Dependency and interdependency may be distinguished analytically, but are used interchangeably in this chapter because relationships between beings involve both in practice. We depend upon nature, but its health depends upon us, as is frighteningly evident. An elderly parent is dependent upon her daughter, but the daughter's well-being is thoroughly affected by that of the parent. Dependency and interdependency are integrated in practice.
4 Citations for Locke (1960) use book and section numbers as is standard. For an exam-ple of moral individualism in action in Locke's era, see McCready (1996).
5 Locke's well-known second treatise of government, which conveys the ideas that he supports, is usually not read in tandem with the first, which rebuts those of Filmer (1991).
6 The head may be "forced to garre cut off some rotten member ... to keepe the rest of the body in integritie: but what state the body can be in, if the head ... be cut off, I leave it to the readers judgement" (James, 1994, p.78).
7 In Filmer's account, Adam was the source of all human flesh, including Eve, and men are the primary force in creating children, so women are naturally subordinate. See *The Anarchy of a Limited or Mixed Monarchy* and *Observations Concerning the Originall of Government* in Filmer (1991).
8 King James aimed to reinforce "The Reciprock and mutuall duetie betwixt a free King and his naturall Subjects" (1994, p. 63) in response to a contractual and hence conditional rendering of monarch–subject bonds.
9 Although Locke uses neither "relational" nor "solidarity," he describes the "Obligation to mutual Love amongst Men" (Locke, 1960, II.5) immediately after introducing the state of nature as "perfect Freedom" and equality, invoking Thomas Hooker's ideas on natural duty.
10 For a fuller analysis of Locke's engagement with patriarchy on this subject, see McCready (unpublished manuscript).
11 Forde (2013) thinks that Locke regarded species distinctions as conventional and nominal, versus essential, so his man/animal and mind/body distinctions, derided as neatly dualist, deserve further thought.
12 This point is also a crux of Okin's (1989) liberal feminist elaboration of Rawls.

13 The original position in Rawls' (1971) social contract theory exemplifies this approach.
14 The central difference between these two is discussed below, as is that between these two and the feminist freedom of Hirshmann (2003).
15 Taylor (1981, 1986) similarly cites all beings' teleological goods and interdependency in relationships, but his bio-centric argument follows liberalism's rational and impartial approach to ethics and employs equality and freedom as criteria of respect for nature.
16 Note that persons are free within natural society in Locke's theory. Freedom precedes the social contract; it is not dependent on it. Moreover, an adversarial orientation and competition are not part of his ontology.

References

Filmer, R. (1991). *Patriarcha and Other Writings*, ed. Johann P. Sommerville. Cambridge: Cambridge University Press.

Fineman, M. A. (2004). *The Autonomy Myth: A Theory of Dependency*. New York: The New Press.

Forde, S. (2013). *Locke, Science, and Politics*. New York: Cambridge University Press.

Held, V. (2006). *The Ethics of Care: Personal, Political, and Global*. New York: Oxford University Press.

Hirschmann, N. J. (2003). *The Subject of Liberty: Toward a Feminist Theory of Freedom*. Princeton: Princeton University Press.

James. (1994). *Political Writings*, ed. Johann P. Sommerville. Cambridge: Cambridge University Press.

Kish, K. (2020). "Reclaiming freedom through prefigurative politics" in *Liberty and the Ecological Crisis: Freedom on a Finite Planet*, eds. Christopher J. Orr, Kaitlin Kish, and Bruce Jennings. New York: Routledge.

Locke, J. (1960). *Two Treatises of Government*, ed. Peter Laslett. Cambridge: Cambridge University Press.

McCready, A. R. (1996). "The ethical individual: An historical alternative to contemporary conceptions of the self." *American Political Science Review*, *90*: 90–102.

McCready, A. R. (unpublished manuscript). "Species comparisons in John Locke's conjugal society."

Merchant, C. (1980). *The Death of Nature: Women, Ecology, and the Scientific Revolution*. New York: Harper Collins.

Okin, S. M. (1989). *Justice, Gender, and the Family*. New York: Basic Books.

Plumwood, V. (1993). *Feminism and the Mastery of Nature*. New York: Routledge.

Quilley, S. (2020). "Liberty in the near Anthropocene: State, market, and livelihood" in *Liberty and the Ecological Crisis: Freedom on a Finite Planet*, eds. Christopher J. Orr, Kaitlin Kish, and Bruce Jennings. New York: Routledge.

Rawls, J. (1971). *A Theory of Justice*. Cambridge, MA: Harvard University Press.

Taylor, P. W. (1981). "The ethics of respect for nature." *Environmental Ethics*, *3*: 197–218.

Taylor, P. W. (1986). *Respect for Nature: A Theory of Environmental Ethics*. Princeton: Princeton University Press.

Tronto, J. C. (1993). *Moral Boundaries: A Political Argument for an Ethic of Care*. New York: Routledge.

6

LIMITS AND LIBERTY IN THE ANTHROPOCENE

Peter F. Cannavò

Introduction

The notion of ecological limits posits a ceiling on the planetary resources and environmental sinks available for production and consumption. The concept, initially popularized by the 1972 Club of Rome report, *The Limits to Growth* (Meadows et al. 1972), has recently been revived with the notion of planetary boundaries (Dobson 2016), i.e. limits beyond which human activity can disrupt various aspects of the Earth system (Rockström et al. 2013). We may also speak about another set of natural limits: the limits of the human constitution, including basic human physiology, mental and emotional needs, mortality, and so forth. Such limits, like ecological limits, involve natural constraints, in this case on what is possible to do with and to human beings.

Natural limits are not uncontroversial. One set of arguments suggests that ecological limits can be overcome through human ingenuity (see below). Libertarians argue that regulations to enforce ecological limits interfere with economic growth, consumer choice, and individual liberty (see Fragnière 2016). There is an additional argument, with more appeal to the political left. This argument, a main focus of critique in this chapter, is that adherence to limits, whether ecological or human, constrains the collective liberty of the political community to govern itself and determine its own fate. Limits straitjacket political deliberation and action, undemocratically turning public life into ecological management by experts. In short, limits are supposedly inconsistent with liberty, whether conceived as liberty from external interference, or liberty as collective democratic self-governance and self-determination.

This chapter does not attempt to make an empirical case for the existence of limits, but considers the implications for liberty of perspectives that either embrace or attempt to deny limits. The central argument is that recognition and embrace of limits, both human and ecological, *enhances* republican liberty as collective self-government among equal citizens. This chapter discusses two political values that connect liberty to limits: *vulnerability* and *non-domination*. Non-domination emerges from the civic republican tradition, while vulnerability has its roots in republicanism and other perspectives, including feminist theory. Together, these two values promote a collective democratic liberty in contrast to the domineering, instrumentalizing impulse to transcend limits. Thus, while the Anthropocene presents stringent ecological constraints as we come up against various planetary boundaries, these limits do not mean that certain forms of liberty cannot flourish. In developing this argument, the chapter critiques two contemporary perspectives, ecomodernism and transhumanism. Both reject natural limits in favour of some notion of collective liberty and self-determination, but they actually end up threatening liberty.

Challenging limits: Ecomodernism and transhumanism

Ecomodernism is a contemporary environmentalist perspective that rejects natural limits. Ecomodernism fundamentally maintains that economic growth can be "decoupled" from environmental degradation (Dobson 2016, p297) through technical ingenuity and the development of substitutes for scarce resources and polluting technologies; ecomodernists also argue that economic growth and prosperity are preconditions for societies to afford environmental regulation. Andrew Dobson (2016) offers empirical arguments to refute these claims. However, ecomodernism can go further and argue against limits as anti-progressive and anti-democratic. This stance focuses on collective liberty and self-determination, rather than individualistic, libertarian values. It is articulated by Ted Nordhaus and Michael Shellenberger, founders of the aptly named Breakthrough Institute. Nordhaus and Shellenberger invoke notions of human progress and collective democratic self-determination, while self-consciously rejecting limits and taking ecomodernism to a teleological level of human perfectionism and dominion over the planet (Nordhaus and Shellenberger 2007). They advocate what they variously term a politics of "overcoming," "self-transcendence," "self-creation," "self-express[ion]," and, in the context not of atomized individualism, but voluntary communities (Ibid., p187, 192, 212). They thus articulate a democratic and communitarian ethos with appeal to political progressives.

Nordhaus and Shellenberger connect this transcendent politics to a rejection of natural limits. They affirm the value of "hubris," (Ibid., p242) and "wealth, power, and self-mastery as virtuous, not evil" (Ibid., p250), and decry environmentalism for its "politics of limits, which seeks to constrain human ambition, aspiration, and power rather than unleash and direct them" (Ibid., p17). An ethic of limits, they argue, prescribes a vision of nature for human society, constrains democratic politics – indeed "claims to be above politics" (Ibid., p145) – and is ultimately authoritarian:

Eco-tragic narratives diagnose human desire, aspiration, and striving to overcome the constraints of our world as illnesses to be cured or sins to be punished. They aim to short-circuit democratic values by establishing Nature as it is understood and interpreted by scientists as the ultimate authority that human societies must obey.

(Ibid., pp131–132)

Nordhaus and Shellenberger characterize human beings as "highly adaptive creatures" (Ibid., p152). Such adaptation includes reengineering the Earth, "remaking nature as we prepare for the future" (Ibid., p113). And, though urging action on climate change, they rather uncritically accept the outcomes of anthropogenic global warming: "As the earth warms, forests disappear, and the Arctic melts into oceans, new natures will emerge all over; it will become increasingly untenable for anyone to claim to represent some essential nature or environment" (Ibid., p239).

Nordhaus and Shellenberger conflate an embrace of natural limits with what John Meyer (2001) calls a *derivative* conception of the relationship between nature and politics. A derivative conception sees the natural world as a guiding authority, embodying blueprints for the substance of political life (pp5–6, 21–34). A derivative view not only relies on controversial, often reductive characterizations of nature, but also has anti-democratic implications, as it potentially empowers an expert elite supposedly equipped to understand and apply the requirements of nature.

Certainly, if one frames natural limits as part of a larger normative order by which human beings must abide, it can indeed involve a derivative relationship between nature and politics. However, limits can merely establish basic ecological conditions or constraints within which politics may operate. Here, limits do not prescribe the substance of politics itself, but merely circumscribe the options, especially economic and consumption options, available to political actors (Fragnière 2016).

Nordhaus and Shellenberger also see a fairly extreme plasticity in nature and, ultimately, humanity itself. Climate adaptation would involve the reengineering of not only the planet, but also human beings:

Global warming is changing nonhuman natures and may end up changing us with it. It will force human societies to adapt in all sorts of ways, not the least of which could be bioengineering ourselves and environments to survive and thrive on an increasingly hot and potentially less hospitable planet.

(Nordhaus and Shellenberger 2007, p253)

This notion of "bioengineering ourselves" in order to "thrive" seems democratic on its face, and ties in with their overall normative vision of individual and collective self-creation and re-creation. However, it actually has troubling implications. More on this below.

The idea of reengineering ourselves is taken to an extreme with transhumanism. Transhumanism is perhaps best associated with Silicon Valley engineers and entrepreneurs seeking physical immortality (Friend 2017); however, it also includes other biological and cybernetic or chemical enhancements or transformations (see, for example, More and Vita-More 2013; Kurzweil 2006; for critiques of transhumanism, see Thomas 2017; Rubin 2014; Fukuyama 2002; McKibben 2003).

One recent transhumanist tract, Steve Fuller and Veronika Lipińska's *Proactionary Imperative* (Fuller and Lipińska 2014), envisions a kind of collective, ostensibly democratic, societal, and human effort to transcend natural and physiological limits. Like Nordhaus and Shellenberger, they envision a combination of individualism and communitarianism. In this case, individuals with similar genetic attributes would self-organize into collectives to subject themselves to experimentation and market the results, though such experimentation would be a kind of civic duty (Fuller and Lipińska 2014, p38, 109, 125, 135). Their version of transhumanism, while admittedly provocative and extreme, brings out the whole movement's disturbing implications. Fuller and Lipińska define transhumanism "as the indefinite promotion of the qualities that have historically distinguished humans from other creatures, which amount to our seemingly endless capacity for self-transcendence, our 'god-like' character" (Ibid., p1). We must overcome our "common natural limitations" (Ibid., p110). They embrace transhumanism in all of its dimensions: human dominion over, and transformation of, the Earth; human mental and bodily modification through genetics or cybernetics; and a complete transcendence of the body by downloading human consciousness into computers.

Importantly, in calling for people to experiment on themselves, Fuller and Lipińska urge that we accept the inevitable suffering, torture, and deaths along the way (2014, p6, 38, 50, 101, 123, 132). They envision perpetual transcendence, with people and nature becoming "means for the cultivation of 'humanity,' understood as a being whose nature is both self- and world-transforming" (Ibid., p37). And transcendence means instrumentalization: nature becomes "disposable waste,"; "the human body itself might be surplus to requirements in an optimally efficient economy – which is to say, one that is fully technologized" (Ibid., p28). Embracing eugenics, they maintain that it emphasizes "the conversion of humanity to capital" (Ibid., p66). Ironically, humanity itself is seemingly exalted but ultimately devalued as mere meat-machine in much transhumanist thinking (McKibben 2003, p203). In the end, the only grounding for Fuller and Lipińska is transcendence itself, which is to say, the endless negation and denial of any sort of meaning or inherent worth.

In sum, the ecomodernists and transhumanists reject natural limits, and not only ecological limits, but also the physiological limits of human beings themselves. In so doing, they embrace an ethic of "overcoming" or endless transcendence. Through a discussion of two values – vulnerability and non-domination – the remainder of this paper will show how this rejection of limits leads to a dangerous instrumentalization of humanity and a denial of liberty, whereas an embrace of limits promotes republican liberty.

Vulnerability

This section will develop the concept of vulnerability and its relation to limits, starting with Michael Sandel's critique of transhumanist enhancement. Sandel, a contemporary republican theorist, acknowledges how the attempt to transcend natural bounds resonates with a sense of mastery over the world and the attraction of freedom from all limits: "There is something appealing, even intoxicating, about a vision of human freedom unfettered by the given" (Sandel 2004). However, the appeal of such limitless freedom is bound up with a desire for power. The ambition to reengineer humanity may be "the ultimate expression of our resolve to see ourselves astride the world, the masters of our nature" (Ibid.).

Sandel argues that this "promise of mastery is flawed. It threatens to banish our appreciation of life as a gift, and to leave us with nothing to affirm or behold outside our own will." Human enhancement "represent[s] the one-sided triumph of willfulness over giftedness, of dominion over reverence, of molding over beholding" (Ibid.). As discussed below, this conceit of mastery involves an instrumentalization of both nature and humanity.

Against the dream of perfection, Sandel opposes the notion of the "the given," "the giftedness of life." While not directly invoking ecological limits, he nevertheless suggests that we must accept and accommodate ourselves to aspects of nature, both human and non-human, that are beyond our control:

> To acknowledge the giftedness of life is to recognize that our talents and powers are not wholly our own doing, despite the effort we expend to develop and to exercise them. It is also to recognize that not everything in the world is open to whatever use we may desire or devise. Appreciating the gifted quality of life constrains the Promethean project and conduces to a certain humility.
>
> *(Ibid.)*

Sandel thus argues, "The awareness that our talents and abilities are not wholly our own doing restrains our tendency toward hubris" (Ibid.).

Sandel's idea of humility in the face of nature is related to a concept of *vulnerability*. "Vulnerability," says John Barry, "connects to the long-standing and foundational green discourse about limits" (2012, p75). But Barry also derives vulnerability partly from the republican tradition. A key theme of republicanism is the limited, vulnerable nature of the polity. J.G.A. Pocock says civic republicanism traditionally regarded a virtuous polity as something that "existed in time, not eternity," and was "therefore transitory and doomed to impermanence" (Pocock 1975, p53). Barry similarly speaks of "the tendency of all naturally occurring and living things (a category that includes the city republic itself) to grow, mature, decline, and eventually die" (2012, p226).

The polity's vulnerability is related to natural constraints, to the vagaries of fortune, and to human imperfection. Republicanism, Barry says, "remind[s] us that our vulnerability to natural disasters and our ultimate dependence on the natural world – and our concomitant dependence on one another – is the fundamental starting point for *any* sort of politics" (Barry 2012, p229). As republicanism has traditionally been sensitive to the polity's environmental conditions, including geography, soil fertility, climate, and population size, Barry sees a republican "acceptance of the (relatively) fixed ecological limits which delimit the material range of within which human society can exist and/or flourish" (Ibid., p219). And vulnerability to nature is related to the contingencies of fortune. In his discussion of fortune, Machiavelli, for example, uses the analogy of a flood-prone river and how it must be restrained and channelled through dikes and dams (Machiavelli 1985, pp98–101). However, Machiavelli sees the river as inevitably rising, with the impacts mitigated but never fully mastered. In other words, neither fortune nor nature can ever be entirely controlled – we must recognize our vulnerability and limitedness in the face of both.

Moreover, a republican polity is vulnerable to corruption, i.e. to the limits of human nature. Given human weaknesses, Pocock argues, republican virtue was bound to fail and yield to corruption, and "the republic, being a work of men's hands, must come to an end in time" (Pocock 1975, p53, 75). Republican theorist Iseult Honohan thus argues, "Political institutions that provide some degree of control over our destinies are still a fragile creation that cannot be taken for granted" (Honohan 2002, p149).

Maintaining a republic against degenerative forces means attention to external limits rather than faith in perpetual transcendence. Republicanism thus avoids a naïve and dangerous optimism that nature, both human and non-human, is infinitely plastic and can be remade to fit the ambitions of those who would reengineer the world.

In fully laying out the notion of vulnerability, Barry (2012), however, goes beyond the republican tradition. Republicanism has tended to emphasize the *collective* vulnerability of the polity. Resonating with feminist and other contemporary moral theories of vulnerability (for example Mackenzie, Rogers, and Dodds 2014; Nussbaum 2006; Butler 2004; MacIntyre 1999), he focuses on the vulnerability of human *individuals*. Accordingly, he says, "to be vulnerable is constitutive of what it means to be human," in other words, that "a human who is invulnerable is not human or at least not comprehensible as human" (Barry 2012, p36). Catriona Mackenzie, Wendy Rogers, and Susan Dodds (2014, p7) describe this as *inherent* vulnerability, "intrinsic to the human condition," and "aris[ing] from our corporeality, our neediness, our dependence on others, and our affective and social natures. We are all inherently vulnerable to hunger, thirst, sleep deprivation, physical harm, emotional hostility, social isolation, and so forth."

Importantly, Barry sees vulnerability as a value rather than a weakness. This point is also developed by feminist theorist Erinn Gilson (2014), who notes that vulnerability, while often stigmatized, actually represents an ontological human condition (p37) that involves openness to others and the world (p130) and "can have positive manifestations and value, enabling the development of empathy, compassion, and community" (p8), rather than being a badge of weakness and incapability (p34). With regard to its connection to openness, vulnerability involves willingness to change and be altered by others (p86), but it also suggests certain fundamental human characteristics that cannot be altered at will. Vulnerability suggests the limits and also the needs of human corporeality (Barry 2012, p39; also Nussbaum 2006; MacIntyre 1999). According to Barry, recognition of vulnerability motivates "care and compassion for [both] human and non-human others and a linked concern for the capacities, institutions, and resources needed by others (human and non-human) for flourishing" (Barry 2012, p75).

Barry says vulnerability is in tension with Enlightenment sensibilities:

> Vulnerability perhaps conveys too much connection with the body, suffering, notions of unchosen limits, and connotations of human dependency on nature, other human beings, or supernatural beings, sufficient to offend the central Enlightenment belief in mastery of both human and non-human nature.
>
> *(Barry 2012, p37)*

Narratives of progress often see vulnerability as an undesirable condition to be transcended: "In the 'Western imaginary' (cultural and normative) vulnerability signifies not just something to be overcome, but also something which was dominant at an earlier stage of human evolution and history" (Ibid., p48). As Gilson (2014) puts it, "The ideal of the invulnerable self is defined by complete self-sufficiency, self-sovereignty and autonomy, independence from others, and an imperviousness to being affected" (p7).

Yet, as Gilson argues, invulnerability is itself a weakness: it is "ignorance of vulnerability" (p7), a "self-deception" (pp86–87). And it is a "willful ignorance" (p88) that promotes oppression, as it enables denial of one's connections with and responsibility toward others, particularly those whom one subordinates and perceives as abject or dehumanized (pp88–82). Invulnerability also involves oppression and domination in a more direct way. It presupposes an ideology of mind/body dualism that justifies control and mastery over the body, other animals, and nature (p83), with women and people of colour often relegated to the subaltern category of "nature" (p85).

By contrast, the recognition of vulnerability means accepting mortality and "abandon[ing] the counterproductive search for control and mastery" (Barry 2012, p66). Recognition of vulnerability can be "an antidote to an 'arrogant humanism' … and sense of superiority over and invulnerability to the non-human world" (Barry 2012, p75). Recognition of vulnerability sets limits to our actions and forces us to recognize thresholds beyond which we undermine human and non-human flourishing (Ibid., p76). That the natural world or human beings can only be pushed so far before they break and that "neither ecological conditions nor the human condition is infinitely malleable" delegitimizes domineering and destructive transformative schemes, though without entailing human passivity: we must go about "finding a mean between these extremes of 'arrogant human-ism' on the one hand and 'ecological quietism' on the other" (Barry 1999, p72).

Circling back to Sandel's criticism of transhumanism, in recognizing vulner-ability we are effectively forced to affirm a certain given-ness or even giftedness in human life and nature, and to adopt a stance of humility rather than mastery. The ambition toward mastery involved in ecomodernism, transhumanism, and the rejection of limits is closely related to the stance of invulnerability that Gilson describes. It promises individual self-creation and collective self-deter-mination and liberty, but in the end instrumentalizes both nature and human beings.

Here, one might respond that an emphasis on vulnerability or humility implies a paternalistic, overly protective approach to politics. This returns us to the initial objection to a politics of limits, namely that it stifles liberty, whether individual or collective. I will answer this objection by showing how vulner-ability promotes non-domination and republican liberty.

Non-domination

Non-domination is an important ideal in republican theory. Individuals enjoy non-domination when they are not subject to the arbitrary will of others and instead enjoy an equal status as citizens and exercise political control over their govern-ment, which is expected to act in their interests (Pettit 1997, 2012; Laborde and Maynor 2008). Republicanism involves a strong emphasis on collective self-government and participation and – relative to liberalism – considerable social and economic equality as hedges against domination.

Philip Pettit (1997, p52) says that "someone dominates or subjugates another, to the extent that 1) they have the capacity to interfere 2) on an arbitrary basis 3) in certain choices that the other is in a position to make." Pettit notes, "An act is perpetrated on an arbitrary basis ... if it is subject just to the *arbitrium*, the decision or judgment, of the agent; the agent was in a position to choose it or not choose it, at their pleasure" (Ibid., p55). "What is required for non-arbitrary state power," Pettit notes, "is that the power be exercised in a way that tracks, not the power-holder's personal welfare or worldview, but rather the welfare and world-view of the public" (Ibid., p56).

However, domination can exist whether or not the capacity to interfere is actually exercised (Pettit 1997, p63). Cecile Laborde and John Maynor note, "For republican thinkers, living in subjection to the will of others *in itself* limits liberty." Thus, "domination is a function of the relationship of unequal power between persons, groups of persons, or agencies of the state." One ought not live at the mercy of another (Laborde and Maynor 2008, pp4–5, emphasis in original). Domination and non-domination are thus status-oriented, political, and relational (Fragnière 2016, p42).

One can interfere without dominating, if that interference "track[s] the interests and ideas" of those affected (Pettit 1997, p65), for example in the case of laws and regulations that protect public health or safety or address collective action problems (Fragnière 2016, p45). However, such interference must be subject to constitutional constraints, checks and balances, public control, deliberation, and contestation (Pettit 1997, p56, 63, 65), as in the case of regulations adopted under the rule of law and through democratic processes. Non-domination itself thus already embodies a notion of limits, though these are limits internal to the political system rather than external, natural limits.

The ideal of non-domination would be in conflict with the programs of ecomodernism and transhumanism. First of all, the reach of these programs – physically transforming the planet and human beings themselves – would be unprecedented. Their limitless, totalizing character means that their impact would enormous and inescapable. In critiquing transhumanism, Bill McKibben conveys this overwhelming, unavoidable impact with the deceptively mundane metaphor of a motorboat:

> Consider paddling around a lake in a canoe: fifty canoes can be exploring the bays and coves and not bothering one another at all. But one motorboat roaring through changes everything for everyone. It may be grand fun for the one guy standing at the wheel with the wind in his hair, but everyone else is left to deal with the wake and the racket and the diesel stink as well as they can.
>
> *(McKibben 2003, p214)*

Unlike the motorboat, however, the sort of projects contemplated by the Breakthrough Institute and transhumanism would have far-reaching and irreversible impacts.

Though advocates of both ecomodernism and transhumanism claim a sort of democratic approach, in practice, some will act as the designers, while others will be the material. The degree of expert proficiency to design and execute these transformations would also encourage the rise of a technocratic elite armed with the capacity for vast, arbitrary power, unbound by any notion of limits, and likely imbued with a conceit of mastery and invulnerability. With ecomodernism and transhumanism, large segments of the population would likely be the unwilling, instrumentalized captives of extreme projects, either because they object to them out of principle or worldview, or because they will be disadvantaged or harmed by them.

Geoengineering carries the potential for collateral human damage, with the impacts inequitably distributed, raising serious justice issues (Gardiner 2010).

For example, solar radiation management, i.e. cooling the Earth by dumping reflective aerosols into the atmosphere, could adversely affect "between 1.2 and 4.1 billion people" through "changes in rainfall patterns" (Shukman 2014). For its part, transhumanism could create extreme hierarchies between the enhanced and the unenhanced, generating pressures on individuals to conform to new standards of augmentation rather than get left behind or subordinated (Thomas 2017). Moreover, the application of physically invasive transhumanist technologies could, as Fuller and Lipińska (2014) acknowledge, involve suffering, death, and the treatment of human beings as raw material: "many in retrospect may turn out to have been used or sacrificed for science" (63). And, as Alexander Thomas (2017) argues, the technologies lend themselves to surveillance and control – Fuller and Lipińska (2014) themselves talk of "mass surveillance and experimentation" (63) – something one already sees in a much more mundane guise with wearable technologies like fitness trackers (Rowland 2019). Thomas (2017) thus warns of the authoritarian nature of transhumanism, as "we may be compelled to conform to a perpetual transcendence that only makes us more efficient at activities demanded by the most powerful system … The ability to serve the system effectively will be the driving force."

By contrast, non-domination would actually thrive through recognition of limits and, more specifically, vulnerability. As noted earlier, one might argue that an emphasis on vulnerability would be paternalistic and anti-democratic. As Gilson argues, however, the potential for paternalism exists when vulnerability is fundamentally regarded as a weakness and as a fixed state inhering in certain groups of people who are considered in some way inferior (Gilson 2014, pp33–38). When vulnerability is recognized as an ontological, inherent condition we all share – without disregarding more specific situational forms of vulnerability (Gilson 2014, p37) – it can take on a more egalitarian, democratic aspect (Barry 2012). Iseult Honohan notes,

> Civic republicanism addresses the problem of freedom among human beings who are necessarily interdependent. As a response it proposes that freedom, political and personal, may be realized through membership of a political community in which those who are mutually vulnerable and share a common fate may jointly be able to exercise some collective direction over their lives.
>
> *(Honohan 2002, p1)*

A shared recognition that we are all vulnerable can be the basis for horizontal, mutual interdependence and collective democratic self-government among citizens, rather than hierarchical dependence and subordination. Moreover, those holding positions of power or influence in government, civil society, and the private sector will more likely have the empathy to ensure that their actions and policies are not arbitrary and that they track the interests and welfare of their fellow citizens.

Recognition of vulnerability also opens up a figurative space free from domination, a space within which one may cultivate democratic empowerment. Recall Barry's observation, quoted earlier, that "neither ecological conditions nor the human condition is infinitely malleable." This suggests that a notion of naturally given, virtually immutable vulnerability, and the associated limits, creates a space that cannot be claimed by domineering, totalizing social, political, or technological ideologies.

Here, one is reminded of a more literal space, namely in Piers H.G. Stephens' discussion of George Orwell's *1984*. Stephens sets the pastoral countryside, to which the main characters periodically escape, against the totalitarian society of Oceania: "the independent regularity of nature offers an epistemological limiting point of sorts to the world-remaking ambitions of a ruling group" (Stephens 2004, p91). He adds, "Nature … functions to protect liberty by curbing the Orwellian excesses of authoritarian definition," including conceits about the infinite malleability of humanity (Ibid., p91).

The idea of human or non-human nature as a limiting condition for human beings' ambitious, transformative schemes not only provides a space free from potential domination, but helps cultivate the ability of citizens to engage in democratic self-government. In his chapter in this volume, Stephens, drawing partly on traditions of agrarian and sylvan liberty, argues that wild or other relatively undeveloped places enable people to interact with settings and entities that have not been marked by other human beings for an instrumental purpose. Stephens argues that such interactions can enable spontaneity, the development of the imagination, and the cultivation of new perspectives (Stephens, this volume). Recognition of natural limits creates such a protected sphere, a safe operating space for both human beings and non-human nature, not only in Stephens' literal sense of green spaces, but also in terms of one's daily existence, identity, and life-activities not being colonized by a totalizing program of relentless change and transformation.

Nevertheless, as Augustin Fragnière (2016) notes, limits do entail restrictions on liberty in two senses: liberty as non-limitation of options and liberty as noninterference by government. The latter is commonly associated with liberalism (Pettit 1997). A recognition of natural limits, both in terms of ecological constraints and the limits of human nature, rules out, even from a purely empirical standpoint, courses of action that try to defy those limits – human beings and the natural world are in fact not infinitely malleable. Moreover, respect for limits involves regulations that entail government interference with destructive or domineering action.

However, the case is different for liberty as non-domination. If individuals' good involves respecting limits and preventing domination by those who would try to violate such limits, then, as Fragnière (2016) argues, members of a political community may empower government to adopt regulations to track those interests. This consideration, coupled with our earlier discussion of vulnerability as involving mutual interdependence, suggests that natural limits are consistent with a collective, democratic form of liberty, indeed the very sort of liberty that ecomodernists and transhumanists contend is inconsistent with limits.

It is also worth noting here that a recognition of ecological and human limits, along with a valuation of vulnerability, does not necessarily collapse into a derivative relationship between nature and politics and crowd out democracy. Participatory democratic institutions can play a key role in developing effective responses to ecological challenges, particularly as they are able to incorporate deliberation, experimentation, and reflexivity in light of local knowledge and changing environmental conditions (Dryzek and Pickering 2019). Moreover, though the validity and importance of natural limits and of related concepts like vulnerability must be recognized, the specification of what those limits and vulnerabilities actually entail is not a purely scientific question, but something that is amenable to democratic deliberation. As John Dryzek and Jonathan Pickering argue, the specification of planetary boundaries and other limits in practice and policy-making can change in light of the turbulent ecological circumstances of the Anthropocene and in light of "political, technological, and economic feasibility." Moreover, "citizens' and policymakers' perceptions about risks to the Earth system are a crucial ingredient in efforts to define planetary boundaries and to shape institutions that could help humanity to stay within the boundaries" (Dryzek and Pickering 2019, p135).

Conclusion

In the Anthropocene, the concept of natural limits, both ecological and human, has become salient, not only as we bump up against or overshoot planetary boundaries, but also as we contemplate technologies that could transform humans at their basic physiological level. Ecomodernists and transhumanists suggest that such limits constrain human liberty and should be transcended if possible. However, a respect for limits recognizes human vulnerability, promotes non-domination, and in fact advances collective, democratic liberty.

References

Barry, J. (1999). *Rethinking Green Politics: Nature, Virtue, and Progress*. London: Sage.
Barry, J. (2012). *The Politics of Actually Existing Unsustainability: Human Flourishing in a Climate-changed, Carbon-constrained World*. Oxford: Oxford University Press.
Butler, J. (2004). *Precarious Life: The Powers of Mourning and Violence*. London: Verso.
Dobson, A. (2016). "Are there limits to limits?" In T. Gabrielson, C. Hall, J. M. Meyer, & D. Schlosberg (eds.), *The Oxford handbook of environmental political theory* (pp. 298–303). Oxford: Oxford University Press.
Dryzek, J. S., & Pickering, J. (2019). *The Politics of the Anthropocene*. Oxford: Oxford University Press.

Fragnière, A. (2016). "Ecological limits and the meaning of freedom: A defense of liberty as non-domination." *De Ethica: A Journal of Philosophical, Theological and Applied Ethics,* *3*(3): 33–49.

Friend, T. (2017, April 3). "Silicon Valley's quest to live forever." *New Yorker, 93*(7). Retrieved from https://www.newyorker.com/magazine/2017/04/03/silicon-valley s-quest-to-live-forever

Fukuyama, F. (2002). *Our Posthuman Future: Consequences of the Biotechnology Revolution.* New York: Farrar, Straus, and Giroux.

Fuller, S., & Lipińska, V. (2014). *The Proactionary Imperative: A Foundation for Transhumanism.* Hampshire: Palgrave Macmillan.

Gardiner, S. M. (2010). "Is 'arming the future' with geoengineering really the lesser evil? Some doubts about the ethics of intentionally manipulating the climate system." In S. M. Gardiner, S. Caney, D. Jamieson, & H. Shue (eds.), *Climate ethics: Essential readings* (pp. 284–312). Oxford: Oxford University Press.

Gilson, E. C. (2014). *The Ethics of Vulnerability: A Feminist Analysis of Social Life and Practice.* New York: Routledge.

Honohan, I. (2002). *Civic Republicanism.* New York: Routledge.

Kurzweil, R. (2006). *The Singularity is Near: When Humans Transcend Biology.* New York: Penguin Books.

Laborde, C., & Maynor, J. (2008). "The republican contribution to contemporary political theory." In C. Laborde & J. Maynor (eds.), *Republicanism and political theory* (pp. 1–28), Malden, MA: Blackwell.

Machiavelli, N. (1985). *The Prince.* H. C. Mansfield, Jr. (Transl.), Chicago: University of Chicago Press.

MacIntyre, A. (1999). *Dependent Rational Animals: Why Human Beings Need the Virtues.* Chicago: Open Court Press.

Mackenzie, C., Rogers, W., & Dodds, S. (2014). "Introduction." In C. Mackenzie, W. Rogers, & S. Dodds (eds.), *Vulnerability: New essays in ethics and feminist philosophy* (pp. 1–29). New York: Oxford University Press.

McKibben, B. (2003). *Enough: Staying Human in an Engineered Age.* New York: Henry Holt.

Meadows, D., et al. (1972). *The Limits to Growth: A report for the Club of Rome's Project on the Predicament of Mankind.* New York: Universe Books.

Meyer, J. M. (2001). *Political Nature: Environmentalism and the Interpretation of Western thought.* Cambridge: The MIT Press.

More, M., & Vita-More, N. (eds.). 2013. *The Transhumanist Reader: Classical and Contemporary Essays on the Science, Technology, and Philosophy of the Human Future.* West Sussex, UK: John Wiley & Sons.

Nordhaus, T., & Shellenberger, M. (2007). *Break through: Why We Can't Leave Saving the Planet to Environmentalists.* Boston: Houghton Mifflin Harcourt.

Nussbaum, M. (2006). *Frontiers of Justice: Disability, Nationality, Species Membership.* Cambridge, MA: Harvard University Press.

Pettit, P. (1997). *Republicanism: A Theory of Freedom and Government.* Oxford: Oxford University Press.

Pettit, P. (2012). *On the People's Terms: A Republican Theory and Model of Democracy.* Cambridge: Cambridge University Press.

Pocock, J. G. A. (1975). *The Machiavellian Moment: Florentine Political thought and the Atlantic Republican Tradition.* Princeton: Princeton University Press.

Rockström, J., et al. (2013). "Safe operating space for humanity." In L. Robin, S. Sörlin, & P. Warde (eds.), *Documents of global change.* New Haven: Yale University Press.

Rowland, C. (2019, February 16). "With fitness trackers in the workplace, bosses can monitor your every step – and possibly more." *Washington Post*. Retrieved from www.washingtonpost.com/business/economy/with-fitness-trackers-in-the-workplace-bosses-can-monitor-your-every-step--and-possibly-more/2019/02/15/75ee0848-2a45-11e9-b011-d8500644dc98_story.html?utm_term=.74eba0ef2384.

Rubin, C. (2014). *Eclipse of Man: Human Extinction and the Meaning of Progress*. New York: Encounter Books.

Shukman, D. (2014, November 26). "Geo-engineering: Climate fixes "could harm billions." *BBC News*. Retrieved from http://www.bbc.com/news/science-environment-30197085.

Sandel, M. J. (2004, April). "The case against perfection: What's wrong with designer children, bionic athletes, and genetic engineering." *The Atlantic Monthly, 293*(3). Retrieved from https://www.theatlantic.com/magazine/archive/2004/04/the-case-against-perfection/302927/

Stephens, P. H. G. (2004, March 1). "Nature and human liberty: The golden country in George Orwell's 1984 and an alternative conception of human freedom." *Organization & Environment, 17*(1): 76–98.

Thomas, A. (2017, July 31). "Super-intelligence and eternal life: Transhumanism's faithful follow it blindly into a future for the elite." *The Conversation*. Retrieved from https://theconversation.com/super-intelligence-and-eternal-life-transhumanisms-faithful-follow-it-blindly-into-a-future-for-the-elite-78538

7

THE VIRTUE ETHICS ALTERNATIVE TO FREEDOM FOR A MUTUALLY BENEFICIAL HUMAN-EARTH RELATIONSHIP

Anna Beresford

Introduction

Developing a mutually beneficial human–Earth relationship requires that humanity adopt the fundamental principle of systems thinking and operate as if everything is connected, "at least weakly," to everything else (Kay and Schneider, 1994, pp5–6). This adoption sounds like a straightforward and simple task, but it reveals a deep tension between our understanding of complex systems and the ontological foundation of the social systems of the Global North. While freedom of the individual is paramount in the democratic societies of the Global North, the understanding of the individual as an autonomous, morally sovereign agent is founded on an operational assumption that there is a dualism of mind and matter, where matter is mechanical and the mind is an objective viewer. Moreover, this cherished concept of freedom on which we operate is entrenched in a philosophical worldview that has led to the "justification" of the overconsumption and subjugation of the non-human world. This ontology of the individual which provides our conception of freedom is at loggerheads with adopting a systems approach, as it requires an individualism that precludes the necessary relationality required for a systems approach. In contrast, I argue that the relational ontology underpinning the *eudaimonian* virtue ethics tradition offers a concept of freedom that is consistent with the relationality required for adopting a systemic understanding of the world. This concept of freedom allows us to operate in a manner that contributes to the mutual flourishing of our human systems and the ecosystems in which they are nested. Likewise, while in some respects the individual may be more constrained by material circumstances, adopting a virtue ethic understanding of freedom does not mean that we forego individual agency; rather, it allows us to live in a manner that is more properly free. I will argue this through a discussion of virtue ethics and its foundational ontology in contrast to our operational ontological understanding of the individual born of Enlightenment thought.

Types of virtue ethics

The last half century has seen a marked resurgence of interest in virtue ethics. This interest is largely attributed to G.E.M. Anscombe's insistence on the futility of doing moral philosophy until we develop an "adequate philosophy of psychology" (1958, p1), because the moral sense – the ought – belongs to an earlier understanding of ethics which has since been left behind. Rather than trying to define the parameters and conditions of moral behaviour, virtue ethics engages with the formation of an individual's virtue and character. Generally, virtue ethics asserts that "facts about virtuous agents and, in particular, facts about virtuous character traits possessed by such agents are more basic than facts about right conduct and are what explain why an action is right or wrong" (Timmons, 2013, p278). Instead of asking: "what is my moral duty or what ought I do?" A virtue ethics approach is to ask: "what would a virtuous person do in this situation?" There are four directions or versions of virtue ethics that occupy contemporary virtue ethic discussions: 1) *eudaimonist* virtue ethics, 2) agent-based and exemplarist virtue ethics, 3) target-centred virtue ethics, and 4) Platonistic virtue ethics (Hursthouse and Pettigrove, 2016). *Eudaimonist* virtue ethics engages with the ontological understanding of the human species rather than with the individual's motivations and, as such, I think it offers the most promising platform for developing an alternate and ecologically responsible concept of freedom. There are various versions of *eudaimonist* virtue ethics: Aristotelian, Platonic, and Stoic. Each has merit in its own right, but I engage with Aristotelian *eudaimonian* virtue ethics through the various versions and revisions from that tradition.[1] Before proceeding with how virtue ethics and its ontological foundation provide an ecologically responsible alternative concept of freedom, I must briefly discuss the ontology of the autonomous individual that needs revision, because virtue ethic's ontology is illuminated by contrast.

The ontology of deontological and consequentialist ethical theories

Deontological and utilitarian ethical theories assume an agent has moral responsibility and culpability, based on the assumption that the individual is free. While commonly understood as ideological adversaries, current deontological and consequentialist ethical theories are built upon a concept of the human person as a rational, objective agent, and engage with distinct concepts such as individual freedom, personal rights, and the parameters that define moral acts. A legacy of the Enlightenment project, this concept of the individual, allows for these ethical theories to grapple with human morality in an abstract manner, regardless of personal motivation or outcome, respectively. In order to separate either personal motivation or outcome, thus enabling a free choice in a moral situation, the individual operates from a dualistic separation of the mind from the body and the rest of the world. There is an implicit hierarchical ordering of the self where rationality and choice reign. Founded on his noumenal/phenomenal distinction, for Kant, moral behaviour must conform to universal principles, and thus material circumstances have no bearing. Likewise, to evaluate only the consequences of a moral act, assumes that we can separate our motivation and do the "right" thing through rational calculation, regardless of our motivations. Both deontological and consequentialist ethical theories rest under the same paradigmatic umbrella, wherein the *possibility* of choice free from restraint is fundamental, and that possibility is built upon an ontological understanding of the world as presented before us.

Whether articulated or not, within this paradigm, the individual effectively considers the self to be a disembodied mind that is rational, autonomous, and free to view the world and choose those things outside of this self. This perspective rests on two distinctions: 1) the distinction of subject and object; 2) the separation of fact and value *or* what "is" versus what "ought" to be. The latter distinction rests on the former. This subject–object distinction is drawn by Descartes in the articulation of his *cogito* (Descartes, 2007 [1637]). As Paul Ricoeur (1974) states, "The philosophical ground on which the *cogito* emerged is the ground of science in particular, but, more generally, it is a mode of understanding in which the existent (*das Seiende*) is put at the disposal of an 'explanatory representation'" (p228). This presents the beginning of the age of "the world as picture." The first presupposition made equates science with research and, in doing so, there is an objectification of what exists and, further, that what exists is placed before the subject. At which point, the subject gains "certainty" of knowledge of what exists under scrutiny. The certitude and representation allow for the *cogito* to be spoken. Then,

> with objectivity comes subjectivity, in the sense that this being-certain of the object is the counterpart of the positing of a subject. So we have both the *positing* of the subject and the *proposition* of the representation. This is the age of the world as view or picture (*Bild*).
>
> *(p228)*

Here we have the objectification of all that exists outside of the mind with this subject/object separation. With this separation, the subject then perceives the self as a rational individual who can view the world as it is presented. With the world before us, to the "disembodied" mind of the individual, "ideas, concepts, and even historical events appear as if passing on a conveyer belt before our mind's eye, from which we take what we consciously decide to make our own" (Zimmerman, 2015, p11). Viewing the world as a picture, the individual primarily identifies as autonomous, rational, and dis-embedded from the rest of the world, free to choose what is presented to the senses. As such, there is a significant amount of personal agency where the individual has the freedom to become anything or anyone, as well as the ability to conform the world to the self's desire. This aspect is made clearer through analysis of the "fact-value" distinction.

The presentation of the world as picture allows for the subject to change the image so that it conforms to how he/she thinks it ought to be, which separates the "is" and the "ought" – a separation of fact and value. This fact–value separation, is present in Descartes' analogy of how he thinks a city ought to be planned, wherein he says,

> these old cities of Europe that have gradually grown from mere villages into large towns are usually less well laid out than the orderly towns that planners lay out as they wish ... they [old cities] make the streets crooked and irregular, you would think they had been placed where they are by chance rather than by the will of thinking men.
>
> *(2007, p5)*

Through objectification of what is before one, as the viewer or planner, one can then mould and change the object or what is, in this case, the city, in any way that one thinks it ought to be done. From this severance we get the exploitation of the non-human world, "justified" through the act of severing the mind and will from the world; the separation of the way things are from the way they ought to be. This separation also leads to the preoccupation in Modern ethics of deriving "ought" from "is." This distinction ultimately results in the denial of the ability to make prescriptive normative statements as argued by Hume (2003 [1738]), and, more recently, by G.E. Moore (2004 [1903]). Such is the operational foundational ontology, derived from Enlightenment philosophical thought, on which our understanding of freedom is built.

Aristotelian virtue ethics and the importance of the *telos*

By contrast, rather than grappling with the moral choices of an autonomous individual, Aristotelian virtue ethics focuses on the development of the virtue and character of the person, engaging with questions such as "what does it mean to live well?" A virtue, according to Aristotle, is "a state of character concerned with choice, lying in a mean, i.e. the mean relative to us, this being determined by reason, and by that reason by which the man of practical wisdom would determine it." (2009 [350 B.C.E.], p31) It is important to note that "mean" does not mean mediocrity, but rather the middle ground between excess and deficiency. For example, courage is the middle ground between rashness and cowardliness, wherein one can achieve excellence. The potential for virtue exists in each of us, but must be cultivated through habit in order to present itself as virtue. Thus acquired virtue can be more simply defined as "a good quality of the mind, by which we live righteously, of which no one can

make bad use" (Aquinas, 2017 [1485]). One is virtuous or lives well depending on whether their actions contribute or detract from the final *telos*, which
in the Aristotelian tradition is *eudaimonia*, commonly translated as flourishing,
well-being, or happiness (2009). The virtue ethics understanding of freedom is
"a condition of the will arising from our nature being in the kind of world that
we inhabit" (Magee, 2015). "Being in the kind of world that we inhabit" factors
heavily with respect to human agency, but that is a point that will be addressed
further on. There are three ontological and foundational assumptions on which
virtue ethics rest: 1) there is a shared human nature and thus a common *telos*
for human beings; 2) facts are value-laden *or* moral statements are factual statements; 3) we are primarily relational beings. Each of these premises is essential
in order to see why virtue ethics provides an alternate, ecologically benign
concept of freedom.

The concept of human *telos* is paramount to the virtue ethics framework.
Central to the pursuit of the good life or living virtuously in the Aristotelian
tradition is "the concept of *man* understood as having an essential nature and an
essential purpose or function." (Aristotle, p58) If there is a common *telos*, then
there must also be a common nature in order for us to be directed towards a
common end. As such, we can either pursue our *telos* well in a manner consistent
with our nature, or act in a way that inhibits our end, detracting from our happiness. Thus, within the Aristotelian tradition,

> 'man' stands to 'good man' as 'watch' stands to 'good watch' or 'farmer' to
> 'good farmer' [...] Aristotle takes it as a starting-point for ethical enquiry
> that the relationship of 'man' to 'living well' is analogous to that of 'harp-
> ist' to 'playing the harp well.'
>
> *(MacIntyre, 2010, p58)*

If the person is acting in a manner consistent with their nature, directing them
to their *telos*, then they are virtuous. Conversely, acting in a manner that detracts
from one's happiness, is not virtuous.

Objections to the concept of telos

This claim will take some unpacking and there are three significant objections
that immediately come to mind. First, one could argue, as Kant (2007 [1781])
does, that the concept of a human *telos* prohibits thought, freedom, and consequently, moral responsibility; basically that a *telos* precludes human agency.
Second, this type of teleological or "purpose" thinking entrenches this ethical
framework firmly in the mechanism from which we are trying to escape. The
last objection is that it commits the naturalistic fallacy and can be used to justify social Darwinism. All of these charges are significant, and would certainly
diminish the appeal of adopting virtue ethics and its consequent concept of freedom as the ecologically responsible alternative.

Kant argues that appealing to a human *telos* prohibits freedom, thought, and moral responsibility. If this is the case, it nulls the value of appealing to virtue ethics for an alternate concept of freedom. He argues that accepting a teleological view of nature requires that there be a common nature, and, if there is a common human nature, it cannot be changing. Individuals exist in a world of flux. But if unchanging nature delineates our purpose and we are in the world of flux they cannot intersect. Therefore, the individual has no agency with respect to his or her end. If we have no choice, we are not free. Furthermore, as thought is so closely tied to freedom, in order to think we must be free. Hence teleology, as it denies freedom, also denies thought. If we are not free, thinking beings, how can we be morally responsible? For how can we say that one has made a bad decision and should be punished if that person cannot even choose? Distilled, the problem is this: teleology precludes agency.

Response to objections

This objection ultimately fails, as it anachronistically assumes the mechanistic view of the natural world ushered in by the early Modern thinkers; such a view was not intended by ancient and scholastic thinkers. In Aristotelian thought the end or purpose of "material substance is *inherent* to it, something it has precisely because of the kind of thing it is by *nature*" (Feser, 2010, p143). This means that, rather than being externally imposed, a *telos* is immanent in the thing by nature of what the thing is. That is, the end or purpose of a being is to act in a manner consistent with what that being is. For example, "that a heart has the function of pumping blood is something true of it simply by virtue of being the kind of material substance it is," (p143) and that would be true regardless of whether or not there was some external mind to denote this purpose. Whereas, with machines, the purpose is imposed externally, thus a *telos* is not inherent in the thing; there is no inherent ability of a clock to tell time. A clock tells time because an external mind has designed it for that purpose. For Kant to deny that there is a human *telos* because it restricts human thought, misinterprets the nature of a *telos* for a living being. The *telos* of a human being is to act in a manner that is consistent with what a human is, which includes the ability to think, deliberate, and choose. Acting in a manner consistent with our human nature directs one to happiness. Thus, to say, as Alasdiar MacIntyre (2010) does, that "every activity, every enquiry, every practice aims at some good; for by 'the good' or 'a good' we mean that at which human beings characteristically aim" (p148), this includes the properties of the mind as well. Humans, as with every living thing, "have a specific nature; and that nature is such that they have certain aims and goals, such that they move by nature towards specific *telos*" (p148). Appealing to a *telos* does not preclude the possibility of agency; rather it is the acknowledgement that there is plasticity within limits. The "limit" in this case being the human species, their essence or nature.[2] In making claims such as "a thing fulfils its end" means that it acts in a manner fitting for its species, or in accordance with its nature,

not that each individual must act in some specific manner or some specific voca-
tion. Very simply, the *telos* of a bunny is to act in a "bunny-like" manner. In this
way, an individual human has the ability to both choose a specific occupation or
lifestyle while still having the capacity to act in accordance with the end – corresponding with one's nature – and fulfil the human *telos*. Thus each individual has
agency, freedom and thought as an individual, but it is the general acknowledgement that there is a common species and that each individual has agency with
respect to those abilities that are possible within the species' limits. Likewise, a
"purpose statement" is not a scientific statement but a metaphysical one.

The coupling of fact and value

As his ethics presupposes his metaphysics, for Aristotle, moral statements are
factual statements (MacIntyre, p59). Thus, inquiring into the nature of human
flourishing by asking what leads to the good life, reveals a radically different
ontological understanding of the individual than the Enlightenment and post-
Enlightenment view of the individual. Framing the investigation in this manner
assumes that one cannot separate value from fact. As MacIntyre states for Aristotle,
"to call something good is therefore to make a factual statement" (p59). This is
because it assumes that morality is based on and consistent with the flourishing of
that individual and the species. Thus, although *eudaimonia* is a moralized concept,
it is one that is grounded in empirical facts. Virtuous behaviour is derived from
the material elements that contribute to our flourishing along with our desires,
capabilities, and interests, inclusive of the intellectual and social aspects of our
species (Hursthouse and Pettigrove, 2016). *Eudaimonia* is a moralized concept and
yet because it is grounded in scientific understanding of the species it does not
fall victim to the naturalistic fallacy. It does not claim that something is good just
because it is seen in nature, but simply, that moral statements do not float free of
facts. Claiming that morality is factual – that is, by not de-coupling fact and value
based on a particular concept of teleology – offers a radically different understanding of morality than the one we have inherited from the Enlightenment, as
illustrated in the debates between deontological and consequentialist ethics.

While virtue ethics is not concerned with defining a universal principle of
morality, such as the categorical imperative (Kant, 1993 [1785]) or the pain/
pleasure maxim of utilitarians (Timmons, 2013), this does not mean that morality is completely relative. The commonality of human nature on which virtue
ethics rests, requires that aspects of morality transcend cultural paradigms. At
the same time, since moral statements are factual ones, this means that morality
is embodied and thus embedded in culture. This embedded nature of morality
means that the expression of moral tenets will vary by culture. MacIntyre states
it this way:

> Aristotle thus sets himself the task of giving an account of the good which
> is at once local and particular – located in and partially defined by the

characteristics of the *polis* – and yet also cosmic and universal. The tension between these poles is felt throughout the argument of the *Ethics*.

(p148)

There are commonalities that contribute to human flourishing, but also the well-being and flourishing of those individuals will look different depending on culture, landscape, time in history, or myriad other factors.

For example, with respect to architectural design, Christopher Alexander (1977, 1979) argues that there is a "timeless way" of building that draws on aspects of the human psyche, which contributes to us feeling at peace and more fully alive. Yet, while there is a "timeless way," the buildings will vary depending on the landscape, available materials, weather conditions, and various other elements. Alexander suggests that there are practices of building that are consistent with our human nature but differ in articulation by landscape and culture; so too with morality in a virtue ethics framework. The embedded morality of virtue ethics that draws on human nature assumes that we are relational beings with a shared nature. Thus elements of moral behaviour transcend culture, but, as moral acts are embodied acts, and morality is embedded in culture, moral behaviour varies in articulation depending on the material circumstances.

Human nature allows virtues to transcend specific cultures while at the same time are manifested in different manners depending on cultural specifics. Aldo Leopold (1986) uses Odysseus' hanging of the slave girls to illustrate how morality evolves, serving his larger point that we need a new bio-centric ethic that is consistent with the evolutionary narrative. Loosely, Leopold's argument is that, upon returning home, Odysseus could in good conscience hang the slave girls because they were outside of the recognized moral community. In contrast, today, we find that act morally repugnant because we have broadened our moral community to (theoretically) include all members of the human species. I disagree with Leopold's interpretation of this event, as I think that it shows the constancy of virtues and their ability to transcend culture as they are founded on human nature. From this, I think Odysseus hanged the slave girls because they *were* a part of the moral community *but* had betrayed his household – an act of treason in ancient Greek society. Thus Odysseus was acting in a manner consistent with the human virtues of fidelity, justice, and hospitality.[3] For if they really were outside of the moral community, why bother to punish them at all? Odysseus' motivation and conscience are largely irrelevant to the issue at hand. However, I think that, rather than illustrating an evolution of morality, this case illustrates the unchanging need to protect the home, and the way that this can differ by expression in culture. Instead of showing the evolving nature of morality, this shows constancy in morality because of the foundation on human nature, *but* that morality is embedded in culture and thus the expression of moral behaviour differs by culture. As it draws on human nature, but allows for various articulations of virtuous behaviour, virtue ethics offers an ecologically responsible alternative concept of freedom.

While this example illustrates the constancy of a human nature, this particular example also helps to illustrate a divergence from Aristotle while remaining within the Aristotelian tradition. In this volume, Jeffery Nicholas in his chapter "Who Stands for Uŋčí Makhá: Liberty, Freedom, and the Voice of Nature," makes two important distinctions regarding this divergence. Nicholas explains how MacIntyre swaps the use of "common good," singular, for "common goods," plural, as Aristotle, "dismisses vast spheres of activity from the good life—the life of the farmer, the life of the home" (Nicholas, this volume). However, adopting the language of "goods" allows for the recognition of a diversity of activities that make up the good life and, in doing so, would prohibit the "just execution" of the slave girls. Second, Nicholas' distinction between "a good" and "Sam's good," prohibits the institution of slavery itself as, while one *could* argue that slavery offers *a* good of sorts, e.g. the exploitation of another for a monetary gain, this cannot be done outside of the context for the good of society itself. As Nicholas states with the example of "Sam," "Sam can identify a good independently from Sam's family, but Sam cannot identify Sam's good independently from Sam's family" (p7). So too with the institution of slavery. While it could be identified (in a perverse way) as a good independent from society, it cannot be identified as a societal good independent from society and is therefore, contra Aristotle, morally impermissible.

Virtue ethics, ecology, and freedom as flourishing

The relational ontology and the embedded morality of virtue ethics make it an ecologically sustainable alternative ethic. Relations are the fulcrum of ecology and systems thinking, both between individual organisms and the nested nature of systems themselves. This type of relational thinking destroys ethical theories that are built upon individualism, rationality, and universal principles, as it undercuts the ontological foundation of this thought. The freedom and choices available to the autonomous, rational individuals are greatly diminished once one has to relinquish or constrain their certainty, autonomy, and individualism. However, the relational ontology of virtue ethics is consistent with ecological theories, as it already assumes that the human is primarily relational and that we can only make choices concerning things relative to us. This is not to suggest that ethics can be subsumed by science, but on the contrary that we can use empirical evidence to inform our ethical behaviour. As moral statements in virtue ethics are factual and, in turn, these are culturally embedded, there is no fundamental transformation needed to the principles of virtue ethics to nest it into larger systems. We simply need to acknowledge that our social systems are embedded in wider ecological systems. This is not a problem for virtue ethics as virtuous behaviour is already developed in a relational and situational manner. There is no major ontological shift required to fit virtue ethics into an era of complexity. But how does this relate to freedom?

The concept of *eudaimonia*, or happiness or flourishing, is central to virtue ethics, but also provides the justification for why virtue ethics offers an ecologically benign alternative understanding of freedom. Freedom, like any condition within the virtue ethics ontology, is relational and subject to the limitations of our nature and the material circumstances in which we find ourselves. Recall, the virtue ethics understanding of freedom is "a condition of the will arising from our nature being in the kind of world that we inhabit" (Magee, 2015). As such, choice, within this philosophical schema, is made through deliberation of means, not ends (Aristotle, p44). For example, Aristotle says that a doctor does not deliberate about whether to heal, but rather, that end is already assumed by nature of being a doctor. As such, what the doctor does deliberate about is the method of healing the patient (p44). The choices that the doctor makes are made within the parameters of what can be done to heal, and, if that can be done through a variety of means, then the doctor deliberates between the methods and chooses the one that seems most appropriate. In this way, the choice is subject to the parameters of the situation – acknowledging that we are a certain type of species and that there are limits with respect to our own bodies, interactions with others, and areas in which we dwell. This means that, as human beings, we are free, but we must act with recognition of the physical limits of our species. In this way, we are free to engage with the material particulars that we encounter. In contrast to the autonomous freedom of post-Enlightenment thought, freedom in virtue ethics acknowledges that we are not completely rational observers survey-ing a world before us and choosing what we please. Instead, freedom operates with the assumption that we are deeply embedded in myriad systems and so are free to act and to choose, but only with respect to those with whom we are in relation as part of larger socio-ecological systems.

Human freedom within the virtue ethics paradigm is re-imagined as the abil-ity to strive for *eudaimonia*, that is, freedom understood as human flourishing. Since the individual of Enlightenment and post-Enlightenment thought is con-sidered rational and autonomous, the individual has the (apparent) luxury of understanding freedom as freedom from constraints, able to choose any option presented. The individual embedded in socio-ecological systems in a world of limits does not have the same options available, but this is consistent with a sys-tems thinking approach as socio-ecological systems are embedded in larger envi-ronmental systems. As Burger et al. (2012) state, "a macroecological perspective on the sustainability of local systems emphasizes their interrelations with the larger systems in which they are embedded, rather than viewing these systems in isolation" (p3). This type of freedom entails that we consider ourselves in a relational way and will limit the *number* of choices available.

Simply because "freedom as flourishing" limits the number of choices does not mean that we are necessarily less free or do without. As relevant today as it was when it was written in 1846, Kierkegaard (2010) opens *The Present Age* with a discussion of thought. He claims that it is endemic of our time that we are paralyzed by thought; even the suicide, "does not die *with* deliberation but *from* deliberation" (p3). The same may be said of the version of freedom that accompanies individualism. In many ways, we are burdened with the number of possible choices available to us, and "to live this war," states Sartre (2001 [1943]), "is to choose myself through it and to choose it through my choice of myself" (p354). By contrast, freedom as flourishing in the virtue ethics paradigm does not present us with the same numeric possibility of choice, because we do not view ourselves primarily or exclusively as individuals. In pre-modern, traditional societies, "individuals inherit a particular space within an interlocking set of social relationships; lacking that space, they are nobody, or at best a stranger or an outcast" (MacIntyre, pp33–34). An individual identifies oneself by being a member of society and is identified that way by others. The relational nature of individuals in this age of ecology is comparable to that of pre-modern, traditional societies. Knowing that we are individuals embedded in society and nested in larger ecological systems means that we do not have the ability to create ourselves from a position of autonomy, but rather that we have inherited a set of relationships by nature of being born. Thus, while we are supposedly "less free" to choose to make ourselves, we do have the freedom to recognize the relationships of which we are already a part, and to act in a manner that is consistent with flourishing within the parameters of those relationships and the wider context of socio-ecological systems.

Let us consider what the above may look like with a more concrete example. A common trope of "sustainability marketing" is to view a system as isolated and closed. Once done, one can then reduce energy use within the "closed system," brand it as a sustainable alternative, and then proceed with business as usual without considering the wider implications for energy flows from the larger systems (Burger et al., 2012, pp2–3). Termed "big brand sustainability" by Dauvergne and Lister (2010) this type of "greening" of corporations can be considered progress towards a sustainable society. However, this option is not available to the virtue ethicist, as an exercise of free choice *must* acknowledge the kind of world that we inhabit. Since we know that we are a part of larger systems we cannot act ethically or freely by ignoring the way that systems operate. In order to flourish, we must act in a manner consistent with the facts that we know about ourselves. This includes acknowledging that we are relational beings, embedded in social systems which, in turn are embedded in wider ecological systems. Thus, to wilfully exclude a systematic approach does not contribute to our flourishing and thus detracts from freely living "the good life." With this understanding, freedom is not considered freedom from constraints. Freedom is re-imagined as the ability to strive for *eudaimonia* or flourishing, acknowledging that we are a specific type of embodied species and that we live in a world of limits.

Perhaps the biggest obstacle to adopting a virtue ethics concept of freedom is that it will require intellectual humility and, in some ways, increased personal responsibility. Acting from a mindset of autonomous individualism, in a sense, one is responsible for creating one's own identity; the individual is the master of one's own life. Likewise, each person has the luxury of certainty with respect to scientific knowledge. However, having to acknowledge that we are relational beings embedded in wider systems, and thus only able to choose things which are relative to us requires intellectual humility and the relinquishing of a controlling mindset (Aristotle, 30). The relational aspect of our choices means that we must forego apparent certainty and live with the knowledge that we may be wrong. With respect to moral decisions, appealing to a universal maxim takes the responsibility from the individual to determine right from wrong. With the relational and embodied virtuous behaviour of virtue ethics, we cannot appeal to a universal principle. Thus, in abolishing certainty and universal principles, we also take on responsibility for our actions in a manner that is not required within a deontological or utilitarian framework. In one sense, this seems unappealing, as it looks like a lessoning of choices and heightening of personal responsibility. On the other hand, it is liberating in a manner that is not possible within the confines of the mechanism and universal principles of Enlightenment and post-Enlightenment thought. We have the responsibility of shaping our character and developing our virtue, but we also have the possibility of expressing virtuous action in as many ways as there are individuals to do so. Moreover, it gives us the freedom to design our societies and express culture in an entirely new ways as we try to transition into a more sustainable future.

Adopting a virtue ethics paradigm does not mean reverting back to some notion of the old days or some idealized version of past culture. Nor does it mean that, because we are drawing upon aspects of an Aristotelian philosophical framework, the laws which he outlines for governing the *polis* are those which we should adopt for the present day. As virtuous behaviour is relational, embedded, and varies in expression from culture to culture, the particulars of a new, virtue-based society have yet to be determined, but will likely look different from any ancient or medieval precursor. A significant reason for this is because we must make use of the current ecological principles and scientific facts in order to theorize about or re-imagine what our new society may look like. Additionally, now that we have had the benefit of living in a "rights"-based society, the health and ease of medical and technical advancements, and availability of career choices, we would be hesitant to relinquish those aspects of life we hold dear. As such, we have to establish exactly what "flourishing" means and looks like in our present age. As Anscombe says, "it is a bit much to swallow that a man in pain and hunger and poor and friendless is 'flourishing', as Aristotle himself admitted" (p15). Likewise, it is difficult to imagine giving up some the options or choices available to us that make our lives comfortable and yet consider this as flourishing. However, it is also quite clear that we cannot call our society "flourishing" if we continue to exploit the people, places, and species with whom we are in relation. The beauty and the difficulty of freedom as flourishing is that it does not offer a rulebook, it leaves us with the responsibility and the opportunity to form our characters and interpret our reality so that we may act in a manner that contributes to the mutual flourishing of ourselves, our societies, and the wider ecological systems in which we are nested.

Notes

1 One of the main differences between Aristotelian, Platonic, and Stoic virtue ethics is that the Aristotelian position argues that virtue is necessary but not sufficient for *eudaimonia*, as it also requires good fortune or luck. Whereas, Platonic and Stoic *eudainmonia* virtue ethics maintain that virtue is a necessary and sufficient condition for the good life. Thus, because Aristotelian virtue ethics focuses on the material circumstances, I think it provides a better platform for engaging with other systems. For a brief but succinct comparison of the different versions of *eudaimonian* virtue ethics, see Rosalind Hursthouse and Glen Pettigrove, "Virtue Ethics," *The Stanford Encyclopedia of Philosophy*. https://plato.stanford.edu/archives/win2016/entries/ethics-virtue/

2 An important distinction must be made with respect to essences or nature. It is not an abstract form or essence that is filled with matter that makes up the nature. Neither is it simply the matter that constitutes the essence. It is both matter and form that constitute the nature where it is the matter existing in the manner that it does. This is an essential contribution that Aquinas made to Aristotelian thought. As Copleston states, "Aquinas [...] while retaining the Aristotelian analyses of substance and accident, form and matter, act and potency, placed the emphasis in his metaphysics, not on 'essence,' on *what* a thing is, but on existence, considered as the act of existing." Essentially, Aquinas' distinction in *De Ente et Essentia*, is that, while for material substances, existence and essence are chronologically simultaneous, ontologically, existence precedes essences and, as such, it is not an empty form that shapes matter, but the way that something exists, as that thing, that makes it what it is. For a full explanation of this, see Aquinas, *On Being and Essence*, 2nd ed., trans. Armand Maurer (Toronto: Pontifical Institute of Mediaeval Studies, 2017).

3 MacIntrye makes a similar note in reference to Penelope and how her virtue of fidelity was similar to and as important to Odysseus as the fidelity of between Achilles and Patroclus, as ancient Greek society depended upon the virtue of fidelity.

References

Alexander, C. (1977). *A Pattern Language: Towns, Buildings, Construction*. New York, NY: Oxford University Press.

Alexander, C. (1979). *The Timeless Way of Building*. New York, NY: Oxford University Press.

Anscombe, G. E. M. (1958). "Modern moral philosophy." *Philosophy*, *33*(124): 1–19.

Aquinas, T. (2017 [1485]). *The Summa Theologiae of St Thomas Aquinas* (2nd ed.). (Fathers of the English Dominican Province, Trans.). Online copyright, Kevin Knight. Retrieved from http://www.newadvent.org/summa/2055.htm#article4

Aristotle. (2009). *The Niocomachean Ethics*. (D. Ross, Trans.). Oxford, UK: Oxford University Press. (Original c. 340 BCE).

Burger et al. (2012). "The macroecology of sustainability." *PLOS Biology*, *10*(6): 1–7.

Dauvergne, P., & Lister, J. (2010). "Big brand sustainability: Governance prospects and environmental limits." *Global Environmental Change*. doi:10.1016/j.gloenvcha.2011.10.007.

Descartes, R. (2007 [1637]). *Discourse on the Method of Rightly Conducting one's Reason and Seeking Truth in the Sciences*. (J. Bennett. Trans.). Retrieved from http://www.earlymoderntexts.com/assets/pdfs/descartes1637.pdf

Feser, E. (2010). "*Teleology*: A shopper's guide." *Philosophia Christi*, *12*(1): 142–159. Retrieved from http://www.jpgociety.org/userfiles/art-Feser%20(Teleology)(1).pdf

Hume, D. (2003 [1738]). *A Treatise of Human Nature*. New York, NY: Dover Publications.

Hursthouse, R., & Pettigrove, G. (2016). "Virtue ethics." In E. N. Zalta (ed.), *The Stanford encyclopedia of philosophy.* Retrieved from https://plato.stanford.edu/archives/win2016/entries/ethics-virtue/

Kant, I. (2007). *Critique of Pure Reason.* (N.K Smith, Trans.). New York, NY: Palgrave MacMillan. (Original work published 1781).

Kant, I. (1993). *Grounding for the Metaphysics of Moral with on a Supposed Right to Lie Because of Philanthropic Concerns* (3rd Ed.). (J. E. Ellington, Trans). Indianapolis, IN: Hackett Publishing Company. (Original work published 1785).

Kay, J. J., & Schneider, E. (1994). "Embracing complexity: The challenge of the ecosystem approach." *Alternatives, 20*(3), 32–39.

Kierkegaard, S. (2010 [1846]). *The Present Age: On the Death of Rebellion.* (A. Dru, Trans.). New York, NY: Harper Perennial.

Leopold, A. (1986). *A Sand County Almanac: With Essays on Conservation from Round River.* New York, NY: Ballantine Books.

MacIntyre, A. (2010). *After Virtue: A Study in Moral Theory.* Notre Dame, IN: University of Notre Dame Press.

Magee, J. M. (2015). "Aquinas and the freedom of the will." *Thomistic Philosophy Page.* Retrieved from http://www.aquinasonline.com/Topics/freewill.html

Moore, G. E. (2004 [1903]). *Principia Ethica.* New York, NY: Dover Publications.

Ricoeur, P. (1974). D. Ihde (Ed.). *The Conflict of Interpretations: Essays in Hermeneutics.* Evanston: University of Illinois Press.

Sartre, J. P. (2001 [1943]). Being and Nothingness. In C. Guignon & D. Pereboom (eds.), *Existentialism Basic Writings.* Indianapolis, IN: Hackett Publishing Company.

Timmons, M. (2013). *Moral Theory: An Introduction* (2nd ed.). New York, NK: Rowman & Littlefield Publishers.

Zimmermann, J. (2015). *Hermeneutics: A Very Short Introduction.* Oxford, UK: Oxford University Press.

8

WHO STANDS FOR UŊČÍ MAKHÁ

The liberal nation-state, racism, freedom, and nature

Jeffery L. Nicholas

Introduction

The radical nature of Alasdair MacIntyre's philosophy has often gone unnoticed. In the 1950s and 1960s, as a Marxist and member of the New Left, MacIntyre was often labelled a traitor (Ali 1972; Anderson 1980; Blackburn 1970) partly, as Paul Blackledge and Neil Davidson (2008b, xvi–xvii) explain, because he chose to publish articles in journals associated with the CIA. His mature work has also been identified with conservativism (Bellioti 1989; Frazer and Lacey 1993, 1994; Kymlicka 2002; Mulhall and Swift 1996; Nussbaum 1989). Much of this literature mistakenly associates MacIntyre with communitarianism. David Ingram, for instance, labels Macintyre a traditionalist conservative (Ingram 1995, 12).

Today, however, the extensive literature on MacIntyre demonstrates that his politics is wrongly associated with conservativism or communitarianism (Knight 2011; McMylor 1994; Nicholas 2012, 2014, 2017); in fact, he has endorsed a reading of his theory as revolutionary (MacIntyre 1998). The revolutionary nature of MacIntyre's politics stems from three of its aspects: (1) the rejection of liberalism, (2) the distinction between practices, institutions, and their pursuit of goods, and (3) a philosophy of the common goods of local communities. These three aspects of his political philosophy embody a notion of freedom that he developed in the 1950s and 1960s. An exploration of MacIntyre's early essays will help us understand freedom as collective self-rule. A community of common goods directs its destiny when each member participates in the rank-ordering of the goods and practices of the community. This approach entails a rejection of liberalism and a subordination of institutions to practices.

An examination of the No-DAPL protests against the Dakota Access Pipeline by Native Americans and their allies shall demonstrate the strength of MacIntyre's criticism of the liberal nation-state as a failure of freedom. These protests, which took place in 2016 at Íŋyaŋwakaǧapi Otpi (Sacred Stone Camp), however, also raise problems for MacIntyre's account of collective self-rule. MacIntyre's understanding of nature and the relevance of his politics to environmental justice remain ambiguous. In contrast, the Lakȟóta beliefs about nature that inspired their organization of the No-DAPL protests force us to ask, who speaks for Uŋčí Makhá? Uŋčí Makhá has value in herself and is an agent in the community; therefore, she must have a voice in the collective self-rule of the community of common goods.

MacIntyre's critique of liberalism

For MacIntyre, the liberal nation-state fails to contribute to human flourishing, both of individual persons and of communities. The politics of the nation-state exclude (1) philosophical analysis, (2) questions about the good life, (3) some forms of life as possible for their citizens, and (4) the possibility of in-depth and logical argumentation in political debate (MacIntyre 1998, 238). These exclusions entail that everyday citizens are prevented from

> engage[ing] together in systematic reasoned debate, designed to arrive at a rationally well-founded common mind on how to answer questions about the relationships of politics to the claims of rival and alternative ways of life, each with its own conceptions of the virtues and the common good.
>
> *(MacIntyre 1998, 239)*

MacIntyre believes the image that best captures the nation-state is that of "a large, complex and often ramshackle set of interlocking institutions, combining none too coherently the ethos of a public utility company with inflated claims to embody ideals of liberty and justice" (MacIntyre 1998, 236). As utility companies, modern nation-states "are governed through a series of compromises" between competing interests of both economic and social kinds (MacIntyre 1999a, 131). Thus, they distribute goods unequally to different interests. Various institutions pursue different goals all the while claiming legitimacy from and access to the power and resources of the nation-state. MacIntyre's analysis of institutions will clarify his argument.

Institutions contrast with practices. A practice comprises a teleologically ordered set of activities aimed at achieving internal goods by adhering to standards of excellence, which extend both human conceptions of the good and human powers to achieve them. In contrast, institutions are necessary to sustain practices because they pursue external goods, which support practitioners in their quest for internal goods. Consider acting as a practice and the local theatre as an institution. An internal good of any practice, such as acting, is highly specific: portraying-a-character-in-a story-so-as-to-better-understand-the-character-from-the-inside.

The local theatre company pursues external goods like funding, a place to house the play, education, and prestige.

For MacIntyre, practices are "always vulnerable to the acquisitiveness [and] competitiveness of the institution ... Without [the virtues] practices could not resist the corrupting power of institutions" (MacIntyre 2007, 194). MacIntyre distinguishes between medieval institutions and modern institutions. Medieval institutions rested on the fact that at least some of the members exercised the virtues. Modern institutions, in contrast, are premised on the belief that they can function well without anyone being virtuous (Beabout 2012, 411–12). Thus, a medieval theatre company functions on the belief that the leaders of the theatre company are virtuous in their pursuit of external goods so as not to undermine the pursuit of internal goods. In contrast, modern theatre companies insist not on virtuous leadership, but on procedural rules in their pursuit of external goods. For MacIntyre, in a society in which virtues were not valued, institutions, but not practices, might flourish (MacIntyre 2007, 193).

In fact, today, modern institutions inhibit the virtues. The importance of the image of the utility company comes through in a critique MacIntyre developed while a "minor participant" in a study of an American electric power industry (MacIntyre 1999b, 321). MacIntyre noticed in this study that social roles seemingly unconsciously determined the kinds of responses individuals gave to questions. In short, individuals compartmentalized their lives. "Compartmentalization" does not name just the fact that individuals occupy different roles in different situations in society. Rather,

> each distinct sphere of social activity comes to have its own role structure governed by its own specific norms in relative independence of other such spheres. Within each sphere those norms dictate which kinds of consideration are to be treated as relevant to decision-making and which are to be excluded.
>
> *(MacIntyre 1999b, 22)*

Modern social structures force most of us to compartmentalize our lives at different points; otherwise they cannot function. Their singular goal is the pursuit of external goods. Compartmentalization assists this pursuit because it alienates the individual from the internal goods of his or her practices.

This compartmentalization conditions the policies of a modern nation-state. Procedurally, it excludes some considerations from decision-making. Then, it forges compromises from its different institutions, each one guided by different considerations in each sphere of its activities. Bereft of philosophical discussion of conceptions of the good life, of deep and logical political debate, the procedures of modern institutional government distribute economic and social goods – external goods – in ways which favour some forms of life and disadvantage others, even while such institutions claim they act neutrally regarding forms of life. Thus, institutions and individuals seek to gain as much as possible while contributing as little as possible.

In this situation, the legitimacy of the ramshackle of institutions that comprise the nation-state hinges on its ability to masquerade as the protector of a nation-wide community. This masquerade legitimizes its authority as the sole user of force. Further, the ideal of a nation-wide community presumes a bond likened to kinship between citizens, but which no "modern, large scale nation-state" can achieve (MacIntyre 1999a, 132). This imposed unity perpetuates a fiction from which others can be excluded or, worse, in which others are harmed because they do not belong. The pursuit of this imaginary bond led to the election of Donald Trump in the United States because of his rhetoric of exclusion.

In contrast to the politics of the liberal nation-state, MacIntyre proposes a politics of the community of common goods.

Collective self-rule

Typically, theories of a politics of the common good speak of the common good in the singular (see Nicholas 2015). Aristotle, the grandfather of this position, writes of the common good in the singular, because he dismisses vast spheres of activity from the good life – the life of the farmer, the life of the home. MacIntyre's concept of practice, in contrast, opens up the cornucopia of activities through which human beings pursue a variety of internal goods: goods of Haŋblečeya, goods of swimming, goods of midwifery. As such, he writes of the common goods of society for a virtuous, flourishing life.

According to this conception of the common good, the identification of my good, of how it is best for me to direct my life, is inseparable from the identification of the common good of the community, of how it is best for that community to direct its life. Such a form of community is by its nature political, that is to say, it is constituted by a type of practice through which other types of practices are ordered, so that individuals may direct themselves towards what is best for them and for the community (MacIntyre 1998, 241).

This paragraph contains two points which we must clarify.

First, the identification of one's own goods are "inseparable" from the identification of the common goods of the community. Typically, we believe that one can identify one's goods regardless of one's relationship with others or one's community. For MacIntyre, however, such an individual identification of one's goods is nonsensical. If, for example, Sam believes Sam's good is to ride bicycles all day long, Sam either may not be able to pursue that good because Sam's family has other goods that they prioritize or may neglect the pursuit of the goods of others, for example, the sanity of Sam's partner who now must earn all the money to support the family as well as Sam's bicycling. Sam can identify a good independently from Sam's family, but Sam cannot identify Sam's good independently from Sam's family.

The second point we must take from this paragraph is MacIntyre's notion of politics. Politics is the ultimate practice of a flourishing community. It consists in all the members of the community rank-ordering the goods of their practices. They do so in order to decide how best to achieve their collective and individual goods.

A community of common goods, when flourishing, has three characteristics. First, the community is one in which members share understandings of goods, virtues, and rules (MacIntyre 1998, 249). These include the precepts of natural law (MacIntyre 1998, 247), which establish the conditions by which members learn from each other (Ibid; MacIntyre 2017, 177–178): such as equality, hearing every voice, and diversity. While these shared understandings might include procedural forms of justice – rules – they do so in conjunction with understandings of goods and virtues. That is, procedural rules in the community of common goods are distinct from that in the liberal nation-state, for they are defined by shared understandings of goods and virtues.

Second, the members of such a community reject the compartmentalization of modernity (MacIntyre 1998, 249, 2017, 264–273). In the community of common goods, the members create a holistic approach to decision-making. They rank-order practices and goods and design institutions to pursue those practices and goods. This provisioning allows each member to pursue his or her flourishing within structures he or she helped to design. Thus, institutions operate, not through the exclusion of virtues and other considerations, but through the communal design of those institutions based on the rank-ordering of practices and goods. As we shall soon see, this ideal embodies MacIntyre's early notion of freedom as collective self-rule.

Third, the flourishing community embraces local markets over national or global markets (MacIntyre 1998, 249). National and global markets impose conditions on communities and individuals, which force agents to compartmentalize their desires and distort the goods they pursue, deny agents access to productive work, and create inequalities that prevent individuals from participating in shared deliberation. National markets might require that members of the community plant or refrain from planting certain crops, forcing them to compartmentalize desires or distort the goods they pursue. Likewise, the institutions of the international market, such as the International Monetary Fund or the World Bank, might impose restrictions on nation-states and their communities regarding some aspects of their economic and social life, as it has done in Greece for example.

A community may exist which does not rank-order the goods of their practices and, thereby, the practices in which they engage. This situation will be unlikely to result in a flourishing community or flourishing members of the community, because the community cannot distribute resources or design institutions so that members can pursue their goods and the goods of the community without competing against each other in those pursuits. For example, Sam pursues bicycling, and Sean art. Neither has considered their relationship or the needs of other members of the community, for example, children. Rather than recognizing their limited resources, including time and the need to care for those dependent on them, they compete for those resources creating a situation in which some, or even all, may be left unfulfilled.

Of course, some communities that do not rank-order their practices and internal goods might flourish in some understanding of the term. They "flourish" through luck or the neglect of others. They may be born wealthy and healthy, or they end up taking from others what they need for flourishing. For a time, so-called first-world nations have been able to "flourish" by extracting wealth from so-called third-world nations. The neo-liberal policies of the US have advanced in the last 40 years now so that one generation – the baby-boomer generation – has flourished, putting the flourishing of the X and Y generations out of reach.

Freedom

MacIntyre's community of common goods captures a notion of freedom as collective self-rule, which he developed while a Marxist and member of the New Left in the United Kingdom in the 1950s and 1960s. Paul Blackledge (2004) contends that three articles – "Notes from the Moral Wilderness," "Breaking the Chains of Reason," and "Freedom and Revolution" – articulate a Marxist project about freedom that MacIntyre came to abandon in 1968. I shall argue, however, that the features of MacIntyre's community of common goods captures the notion of freedom found in these essays.

These three essays are centred around the notion of freedom and the discovery of desire. Articulating a Marxist position that rejects Stalinism and liberalism, MacIntyre contends that the goal is to unite fact – or persistent human desires – and value. Thus, MacIntyre takes issue with the fact–value divide that characterizes much of modern moral theory. (For more discussion of this point, see Beresford, this volume.) Both Stalinism and liberalism divorce persistent human desires from value. For Stalinists, the end justifies the means; mass murder and starvation are justified because they shall bring about the socialist state. The moral ought – the socialist state – is divorced completely from the desires of human life, such as for life and food. For liberalism, moral principles are autonomous from desire because they have no non-moral basis. Thus, the moral ought – equality, for instance – is divorced completely from the desires of human life.

The history of class struggle, then, in both the Stalinist and the liberal nation-state, is the history of the denial of persistent human desire in various forms. Reflecting on persistent human desire, reason allows human beings to discover possibility and to frame purpose – the moral ought. "Men are understood not in terms of that which they have been but in terms of the intersection of what they have been and what they can be" (MacIntyre 1960a, 200). Flourishing human beings – what we can be – are able to direct their lives according to their desires. For MacIntyre, freedom rests with a discovery, grounded in history, "of the kind of life in which fundamental desires intentions, and choices are made most effective, in which man [sic] is most agent and least victim" (MacIntyre 1960b, 21). This discovery entails that human beings become aware of themselves as working for freedom, which can only occur through a consciousness of class and engagement for change.

Blackledge identifies "Freedom and Revolution" as the most substantive of these essays in part because MacIntyre articulates a vision of Marxist freedom that contrasts with the bourgeois notion by examining the relationship between the individual and community.

> Because the individual exists in his social relations and because the collective is a society of individuals, the problem of freedom is not the problem of the individual against society but the problem of what sort of society we want and what sort of individuals we want to be.
>
> *(MacIntyre 1960b, 23)*

What kind of society do we want and what kind of individuals do we want to be are questions that point to persistent human desires. We answer the question, "what do I really want?" by joining with others to uncover persistent human desires. Thus, freedom rests in the dialectical relation of individual and community. This relation requires "discovering and making a common shared humanity" (MacIntyre 1959, 94) and the discovery that what people most desire are "certain ways of sharing human life" (MacIntyre 1959, 95).

This notion of freedom as collective self-rule grounded in the unity of fact and value defines the features of the community of common goods examined above: shared understandings, a rejection of compartmentalization, and local markets.

First, the discovery of a common shared humanity is just the discovery of shared understandings. For MacIntyre, these shared understandings are not something static and unchanging. They are always open to questioning as we discover persistent human desires. Dissenting voices are necessary for this discovery. The precepts of natural law ensure that persons in the community are able to contribute to the debate on these shared understandings. "What will be important to such a society ... will be to ask what can be learned from such dissenters" (MacIntyre 1998, 251; confer Nicholas 2012, 171–217).

Second, because they share a vision of the community, and of what kind of society they wish to be, members of the community also resist compartmentalization. The institutions of the society function so as to ensure that members of the community rank-order their practices and the goods of those practices. They make their desires most effective by rejecting compartmentalization, for compartmentalization requires that some desires are not only not effective, but are not considered in decision-making.

Third, freedom is not conceived as the individual against society as depicted in liberalism and market capitalism. Market capitalism pits the individual against society both so that the capitalist can pursue unchecked accumulation of surplus value, and so that individuals compete for goods and for selling their labour in ways which undermine the unity of desire that could undermine the market. Local markets, in contrast, support the rank-ordering of desire and ways of sharing human life. Freedom comprises the continual creation of society, its practices and institutions through a politics in which all members rank-order practices and goods.

I disagree, then, with Blackledge's claim that MacIntyre abandoned the notion of freedom he developed in his Marxist phase. Rather, MacIntyre focused instead on the divorce of fact and value, which he came to label emotivism, because this divorce threatens human flourishing. Further, he switched from discussing facts as persistent human desires to discussing goods, the language of Thomistic-Aristotelianism. These changes were mistakes; both by focusing on the fact–value divide and changing his language, MacIntyre obscured the notion of freedom that is at the centre of the community of common goods. More importantly, however, this focus on the divorce between fact and value and the discussion of goods has prevented MacIntyre from extending his analysis of freedom as collective self-rule through the collective discovery of persistent human desires as he might have done. Yet, the collective self-rule of a community of common goods embodies the notion of freedom in which human agents make most effective their desires, the core of his Marxist notion of freedom.

The No DAPL protests demonstrate the strength of MacIntyre's analysis of nation-state liberalism as a lack of freedom understood as collective self-rule.

No DAPL as a critique of liberalism

The protests over the Dakota Access Pipeline exemplify how those with political and economic power stand against citizens directing their lives through rank-ordering their goods. The pipeline sought to transfer oil from the Bakken Oil Fields to southern Illinois. Originally, the pipeline was projected to run near Bismarck, North Dakota. Bismarck objected to this path because the pipeline ran too close to the water supply for its residents. The new path of the pipeline took it under the Missouri and Mississippi Rivers near Standing Rock Indian Reservation (Dalrymple, August 18, 2016). Where Bismarck supports a population of 72,000, the Standing Rock Indian Reservation supports a population of less than 10,000. These facts show a competition over goods – where does the pipeline go and who retains safe drinking water – and depict the government as a utility company which serves the greatest interest.

Two other issues, however, prove more important than population in relation to freedom. First, the pipeline's path would disturb the sacred burial grounds of Native Americans. Second, the environmental risks to the Missouri and Mississippi Rivers was not properly vetted before the approval of the pipeline. The fact that the interests of Native Americans in their sacred burial ground lost to the interests of Bismarck raises issues of racial justice and imperialism. The fact that the environmental study was inadequate shows that nature has no voice in policy decisions. In both cases, persistent human desires – for honouring the dead and for a clean environment – were ignored as the liberal nation-state sought to satisfy the most customers. However, the sins of the liberal nation-state do not end there.

An elder of the Standing Rock Lakȟóta, LaDonna Brave Bull Allard, organized a cultural celebration and protest camp (Liu, September 13, 2016). Thousands gathered to peacefully protest the pipeline and to educate themselves and others about environmental justice and the harms of fossil fuels. In response, Energy Partners hired TigerSwan, a security firm, to protect their interests over the pipeline. The ex-military security guards of TigerSwan physically and mentally assaulted peaceful protestors at Standing Rock. Leaked documents show that TigerSwan labelled the peaceful protestors as jihadists and worked with state and federal police using "military style counterterrorism measures" against the protestors (Brown, 27 May 2017).

This response captures all of the elements MacIntyre criticized in the liberal nation-state. At no point did the nation engage in (1) a philosophical discussion of the values involved, (2) questions of the good life, (3) recognition of the exclusion of some forms of life, or (4) in-depth and logical argument about the protests. The media barely covered the issue, much less did it provide deep and logical argumentation about the claims involved (Naureckas 2017). At no point did the people of TransCanada sit down with the people of Bismarck and Standing Rock to engage in a systematic, reasoned debate.

Further, the nation-state acted as a composition of ramshackle institutions. Here, the local government of Bismarck, the interests of Energy Partners and their allies, the military which trained the members of TigerSwan, TigerSwan itself, and the tribal system in the US compete for goods. This brief account testifies that those with political and economic power impose their ordering and pursuit of goods over others. Despite thousands at the protests and national sympathy protests, the pipeline is operational today. Then-President Barack Obama intervened late in the game. Near the end of his presidential term, in December 2016, he halted the construction of the pipeline and ordered the Army Corps of Engineers to conduct a new envionrmental impact survey. Obama knew this order to reassess the environmental factors was not strong enough to last once a new president was in power a month later. Moreover, he knew that Energy Partners were advanced far enough that TransCanada, the owner of the pipeline, was willing to pay low fines to finish their work: "As stated all along, ETP and SXL fully expect to complete construction of the pipeline without any additional rerouting in and around Lake Oahe. Nothing [the Obama] Administration has done today changes that in any way" (Business Wire). At best, Obama acted to preserve his legacy as a progressive president.

Obama's actions demonstrate the reality of the liberal nation-state as a utility company. Relative to the Lakȟóta and other occupants of Standing Rock, the city of Bismarck exerted their will to move the pipeline because they have greater political and economic power. This power results in part because of a history of American treatment of Native Americans. Because legitimacy of the state hinges on its ability to use force, that state and federal police worked with

TigerSwan gives this ex-military security firm legitimacy. This legitimacy rested on excluding Native Americans from decision-making as jihadists. Obama had continued to speak out against police violence against black Americans. In those situations, however, profit was not in jeopardy. He failed to speak out opposing the violence against Native Americans by TigerSwan and state police forces. The liberal nation-state systematically prejudices some ways of life – the pursuit of profit – over others – the honouring of the dead.

In approving the Dakota Access Pipeline, Obama's successor, Donald Trump said,

> From now on, we're going to be making pipeline in the United States. We build the pipelines, we want to build the pipe. We're going to put a lot of workers, a lot of skilled workers, back to work. We will build our own pipeline, we will build our own pipes, like we used to in the old days.
>
> *(Mufson and Eilperin 24 January 2017)*

Nationalism goes hand in hand with capitalism, and individuals in the liberal nation-state can only see the government as a utility company that provides scarce goods for which they are in competition. Under capitalism, the persistent human desire for a clean environment is denied. It is denied because people need jobs that environmental protections stop from being created. As predicted, the pipeline has already poisoned the environment (Brown 9 January 2018).

While MacIntyre's analysis of the liberal nation-state as a utility company that uses force to exclude some ways of life and promote others proves true, it seems to offer little in the way of the primary issues surrounding No DAPL. Those issues include racism and the environment. Uncovering the links between racism and the domination of nature will help us extend MacIntyre's idea of freedom to include nature.

Racism, the domination of nature, and the liberal nation-state

MacIntyre's analysis at best implies a relationship between nationalism and race, on the one hand, and the function of the liberal nation-state as a utility company that distributes external goods disproportionately on the other. For Max Horkheimer and Theodor Adorno, in contrast, race functions to support capital accumulation. It does so by hiding capitalist accumulation as production. Horkheimer and Adorno claim that "Bourgeois anti-Semitism has a specific economic purpose: to conceal domination in production" (141). They relate a telling example.

According to Horkheimer and Adorno, the capitalist claims that he produces value, but he does not. Yet, the capitalist is able to hide this fact through racism. "That is why people shout: 'Stop Thief!'—and point at the Jew. He is indeed the scapegoat" (Horkheimer and Adorno, 142). Historically in Europe, the Jew was often a scapegoat for the rich who could say that the Jews who provided loans were the real thieves. This scapegoating lent support to Hitler's fascist takeover of the Weimar Republic and the extermination of millions of Jews.

The history of the US reveals a similar racist attitude that hides capital production. This history is one of the US making peace treaties with Native Americans and then stealing the land ceded in treaties. For example, the Fort Laramie Treaty of 1868 between the US and the Lakȟóta gave Pahá Sápa (the Black Hills) to the Lakȟóta in perpetuity. Pahá Sápa is sacred to the Lakȟóta; it represents the heart of all that is. Once gold was discovered in Pahá Sápa, however, the US Army violated the treaty to "protect" gold miners. This violation led to further wars and the defeat of the Lakȟóta. Yet, just as in Germany someone might yell "Stop Thief!" and point at a Jew, in the US, we tell people not to be an "Indian giver" – someone who gives a gift and then takes it back. Race hides the fact of capital accumulation. In the US context, this capital accumulation lies, not just in the search for gold in Pahá Sápa, but in the initial theft of land from Native Americans. To add insult to injury, the US carved the faces of four presidents into Tunkasila Sakpe ("The Six Grandfathers," what Americans call Mount Rushmore), which sits in the middle of Pahá Sápa. Each of these presidents contributed to violence against Native Americans. Thus, this sacred territory was not only stolen from Lakȟóta, but desecrated with the faces of those who harmed them.

> For capitalism must justify and mystify the contradictions built into its social relations—the promise of freedom versus the reality of widespread coercion, and the promise of prosperity versus the reality of widespread penury—by denigrating the 'nature' of those it exploits: women, colonial subjects, the descendants of African slaves, the immigrants displaced by globalization.
>
> *(Federici 2014, 17)*

TigerSwan's use of the term "jihadists" for the protestors ties together anti-Semitism and racial hate against Native Americans. "Race … is a regression to nature as mere violence … Race today is the self-assertion of the bourgeois individual, integrated into the barbaric collective" (Horkheimer and Adorno, 138). For Horkheimer and Adorno, the bourgeois individual, who is supposed to be the pinnacle of Enlightenment thought, is in fact barbaric. Prefiguring Marxists scholars who come after them (e.g. Silvia Federici and Maria Mies), Horkheimer and Adorno see that capitalism and capitalist production is intimately connected to the human domination of nature (confer Nicholas 2012). To extract value from nature, capitalism depends on devaluing nature as an empty vessel for human exploitation. In dominating nature, the capitalist asserts his or her identity as barbaric.

MacIntyre's discussion of collective self-rule in a community of common goods recognizes the way that race can support the exploitation of some for the benefit of others; yet, his analysis falls short of recognizing that race masks capitalist production as a domination of nature. However, a further exploration of practices and of the Lakȟóta tradition opens a way to expand the idea of freedom as collective self-rule to include nature as a participant in a community of common goods.

Haŋbléčeya and the Sacred Hoop

To develop the promised insights, I will contrast two practices MacIntyre discusses with the Lakȟóta practice of Haŋbléčeya.

MacIntyre's ideal of the fishing community contrasts with his understanding of the liberal nation-state. People might begin fishing "for purely economic reasons" but "discover that their lives and their livelihood now depend on other people, in whom they have to put their trust, and that those other people depend on them not only to do their work well ... but also expect them to be prepared to risk their lives on occasion to save other crew members" (MacIntyre 2011, 18). Fishing is dangerous, a practice which requires virtues. For instance, fishers rely on each other to be willing to rescue boats that are in trouble at sea. In the fishing community, individuals discover that their livelihoods depend on each other and must be willing to sacrifice their lives for others, because they too may someday need help from them. In contrast, in the liberal nation-state, people must be convinced to sacrifice their lives through the belief in nationalism and a national identity opposed to outsiders.

Yet, "[a]lmost everywhere fishing crews are now suffering badly, as a result of overfishing, as a result of forces that make it difficult to survive, unless you participate in large-scale factory fishing" (MacIntyre 2011, 17–18). Where success for fishing crews includes a shared goal "not to overfish the accessible fishing grounds, so as to deprive themselves of their livelihood" (MacIntyre 1991, 71), fishing crews have failed. Here, fish are a means to an end – the livelihood of the fishers and their communities. To maintain that practice and the good of the way of life of fishing as sustainable, fishers must renew the sea. This need for renewal contrasts with the capitalist economy that drives large-scale factory fishing.

What place do fish have in this analysis?

To answer this question, consider MacIntyre's comments on farming:

> on a particular farm some of those cooperating in the farm work may be acting primarily or only for the sake of the specific goods internal to farming, that is, so as to be excellent *qua* farmers in respect of such goods as the renewal of the earth, the living out of the cycles imposed by nature with respect for nature.
>
> *(MacIntyre 1985, 242)*

Here, the earth must be renewed; excellent farmers live out a certain kind of life that respects the cycles imposed by nature. Yet, "[a]n internal good is not an end-product of its activity; it is achieved in carrying the activity through to a successful completion" (1978, 10–31). As a clarification, he states that agriculture, in addition to its internal goods, produces turnips and money. For MacIntyre, turnips comprise, not an internal good of a practice, but an external good. Likewise, fish provide external goods for fishing crews: food, money, and other things one might make with fish (e.g. needles from fish-bones). Turnips and fish seemingly have no value in themselves and are means to an end.

What does this view entail for a community of common goods in which "fundamental desires, intentions, and choices are made most effective, in which man is most agent and least victim?" (Freedom and Revolution, 21) A community exercises agency through collective discovery of their desires as goods in practices and in rank-ordering their practices and the goods of the practices. Fishers and farmers can articulate their desires, intentions, and choices in a way that makes them most effective. Engaging with others to rank-order the practices – fishing and farming, among others – and internal goods of practices, fishers and farmers participate in collective self-rule. They also establish conditions for the structure of institutions in their community. Local markets will not require overfishing the seas and will help farmers to live according to the cycle of nature, producing turnips in the summer and asparagus in the winter. Yet, establishing the institutions of societies allows fishers and farmer, but not fish and turnips, to participate in collective self-rule.

Nature is necessarily excluded from the community. Such a community rests on particular European-American views of nature and the human relationship to it. God created the earth and gave it to human beings to tend their needs. In some versions of the tradition, human beings are called to dominate and tame nature, while in others human beings are called to steward nature. Neither of these situations call for giving nature a voice in collective self-rule.

Contrast fishing and farming and the implied understanding of the value of nature in them with that of the Lakȟóta tradition and their practice of Haŋbléčeya.

Haŋbléčeya, or crying (lamenting) for a vision, involves a number of activities: a sweat lodge, smoking a pipe with a wičháša wakȟáŋ (holy man), taking the pipe into the wilderness to smoke it and wait for the vision. Typically, the individual wears only a specially made blanket during Haŋbléčeya and does not eat. Only the worthy receive a vision. If one receives a vision, then one must consult again the wičháša wakȟáŋ to interpret the vision.

According to the Atka Lakota Museum and Cultural Center, "The most important reason for the Vision Quest is so a person can understand better his/her oneness with all things and gain knowledge of the Great Spirit." ("Haŋbléčeya") Given this function, one good of the practice of Haŋbléčeya is understanding-one's-oneness-with-all-things-through-Haŋbléčeya. This practice incorporates Lakȟóta beliefs about nature. One must understand one's relationship to all things, which includes nature and its entities. The Lakȟóta who practices Haŋblečeya cannot be ambiguous about his or her relationship to nature, because the practice entails coming to an understanding of that relationship.

If a person wants to understand the good of human life and the goods she pursues, she must become familiar with the debates about the nature of the good in her tradition(s). Traditions, for MacIntyre, are socially and historically embedded arguments about the good and other fundamental agreements (MacIntyre 2007, 222; see also 204–225; MacIntyre 1988; Nicholas 2012). Lakȟóta consider nature as a good in itself independently of their subjective interests. Lakȟóta represent their relationship to the earth through the image of the sacred hoop (Nicholas 2012, 207–210; see also, Hoffman 1997; Holly 1994). The sacred hoop is a circle that represents the interconnectedness and dependency of all things on each other. The buffalo is sacred, just as the human person is sacred. They do not comprise objects for the mere utility of human beings.

Vine Deloria Jr., a Native American and professor of political science, helps us to understand the internal good of Haŋblečeya and the Lakȟóta tradition. He reports that Tȟáȟča Hušté (Lame Deer), a Lakȟóta wičháša wakȟáŋ and a leader of the American Indian Movement, held that all creatures have their own fates. *Wakȟáŋ Tȟáŋka* (the Great Spirit)

> only sketches out the path of life roughly for all the creatures on earth, shows them where to go, where to arrive at, but leaves them to find their own way according to their nature, to the urges in each of them.
>
> *(Kindle loc. 589)*

All creatures are, not only created by *Wakȟáŋ Tȟáŋka*, but also have a certain agency. They must find their way according to their nature and their natural urges.

Deloria continues. "We are then, in a real sense, cocreators with the ultimate powers of the universe because in striving to fulfill our destiny, we make the changes that help spiritual ideas become incarnate in the flesh" (Deloria, Kindle loc. 602). The Lakȟóta believe that *wakȟáŋ* imbues life, is spread throughout creation. A particular forest might be *wakȟáŋ*, or a wolf might be *wakȟáŋ*. If we are to renew the earth, we must be respectful of its *wakȟáŋ*. Moreover, if we are to be free, we must find some way to allow different animals and different parts of nature to speak to us, to find some way to listen to them, and to include them in discovering our desires together and the potentialities for our future.

The Lakȟóta view of nature, then, has implications for a community of common goods that contrast with that given of MacIntyre above. Nature – the particular environment of this or that community – is a member of the community on this Lakȟóta perspective. It is imbued with *wakȟáŋ*, having its own desires and destiny. Fish and turnips, buffalo and wolves, do not exist only for the subjective purposes of human beings. If we are to be true to this vision of nature, then we must invite nature into collective self–rule.

How different communities of common goods approach this issue practically must be left to them. Deloria writes, "we no longer depend on the presence and wisdom of elders who can consult with the spirits and give us their counsel when making important decisions. Most of us cannot even fathom how living in that manner would be" (Kindle loc. 377). Yet, if nature is *wakȟáŋ*, we must make space for it in our collective decision-making. To deny it this place in collective decision-making is "to reduce nature to mere violence." Whatever approach is adopted, however, must have real force, in contrast to the environmental studies performed by the Army Corp of Engineers for the Dakota Access Pipeline. This force might mean that environmental studies have final veto or that members of the community stand for nature and have veto over policies.

Conclusion

In the 1950s and 1960s, MacIntyre developed a notion of freedom that rested on the discovery of the unity of the moral ought and facts. This vision of freedom cannot be found in the liberal nation-state which comprises randomly inter-locking institutions competing for common goods. To realize this freedom, political communities need to have philosophical discussions, engage in ques-tions about the good life, allow for citizens to pursue those forms of life they flourish in, and embrace deep and logical argumentation. The No DAPL pro-tests highlight the failings of nation-state liberalism. Yet, it does more, because it uncovers the roots of such liberalism – the denial of the unity of fact and value – in racism as a disguise for capitalist production. Such capitalist produc-tion necessitates a domination of nature. Nature is effectively excluded from the liberal nation-state, and Lakȟóta ways of life are squashed under the capitalist machine.

However, MacIntyre's ideas about freedom and his analysis of the liberal nation-state fail to illuminate the relationship between the divorce of fact and value, racism, and the domination of nature. The Lakȟóta practice of HaŋbleČeya and their tradition underscore what is missing from MacIntyre's analysis: an understanding of nature and the environment as sacred. Filled with *wakȟáŋ*, nature is an equal member of the community of common goods. As such, it deserves a seat in the rank-ordering of practices and common goods. Unable to speak for itself in such deliberations, I recommend that someone stand for Uŋčí Makhá.

The Roman Catholic tradition, which MacIntyre identifies with, might have room for recognizing the value of non-human species and the value of nature. Pope Francis, for instance, echoes Deloria.

> The ultimate purpose of other creatures is not to be found in us. Rather, all creatures are moving forward with us and through us towards a com-mon point of arrival, which is God, in that transcendent fullness where the risen Christ embraces and illumines all things. Human beings, endowed

with intelligence and love ... are called to lead all creatures back to their Creator.

(2015, 83)

While much separates the theology of Roman Catholics from that of the Lakȟóta, this overlap deserves further research. Further, St. Francis of Assisi, who preached to the animals, might be a way of driving this conversation forward so as to recognize, not just the dignity of non-human species, but that such dignity entails that someone stand for Uŋčí Makhá.

To be fair, MacIntyre recognizes that non-human animal species, like dolphins, flourish in their particular way according to their species (MacIntyre 1999a, 24). In Anna Beresford's contribution to this volume, "The virtue ethics alternative to freedom for a mutually beneficial human–earth relationship," she argues that an Aristotelian teleological virtue ethics can support a better human–earth relationship. Her reading of MacIntyre regarding the flourishing of members of species echoes mine. This recognition is an advance over traditional European-origin philosophy, because it renews insights stemming from Aristotle. Aristotle, for instance, believed that all living beings have a soul. Yet, we might wonder how MacIntyre can reconcile his position about dolphins and wolves with his position about fish, at least, if not also turnips. One notable distinction, of course, that Beresford does not mention is that, as far as we know, only the flourishing of human beings involves virtues. Since virtue is a matter of deliberation, as she rightly notes, non-human species cannot exercise virtues in the requisite since. For our understanding of freedom as collective self-rule, MacIntyre's arguments about the flourishing of non-human species seem tangential at best and do not touch on the argument I have made herein. Nothing Beresford writes about the issues contradicts my reading of MacIntyre regarding turnips and fish.

The arguments I have made here regarding the nature of the community of common goods and the rejection of the liberal nation-state have only brushed past virtue ethics. Recall that MacIntyre distinguishes between medieval institutions which presumed virtuous agents and modern institutions which do not but rely, instead, on proceduralism. Beresford contrasts deontological and consequentialist ethics with their reliance on an ontology of radical individualism with a virtue ethics ontology of relationality to defend her version of virtue ethics. In doing so, she has developed the features of virtue ethics that would support people in free communities of common goods. Members of these communities must see themselves relationally – as noted with MacIntyre's claim that one cannot identify one's own good without first identifying the goods of the community in collective self-rule. Where Beresford defends the need to consider the larger systems to which the community belongs as a limit on their collective self-rule, I argue that such self-rule entails that someone speak for Uŋčí Makhá. This standing for Uŋčí Makhá mirrors Beresford's conclusion regarding virtue ethics for human–earth relationships. Beresford quotes Burger: "As Burger et al. state, 'a macroecological perspective on the sustainability of local systems emphasizes their interrelations with the larger systems in which they are embedded, rather than viewing these systems in isolation'" (Burger et al. 2012). She adds: "this type of freedom, that entails that we consider ourselves in a relational way, will limit the *number* of "ethical" choices available" (Burger et al. 2012). I contend that Uŋčí Makhá is a member of the community of common goods who participates in rank-ordering the goods of the community. At the least, she should be able to limit the number of "ethical" choices considered.

References

Ali, T. (1972). *The Coming British Revolution*. London: Jonathan Cape.

Anderson, P. (1980). *Arguments within English Marxism*. London: Verso.

Beabout, G. (2012). "Management as a domain-related practice that requires and develops practical wisdom." *Business Ethics Quarterly, 22*(2): 405–432.

Belliotti, R. (1989). "Radical Politics and Nonfoundational Morality," *International Philosophical Quarterly, 29*(1): 33–51.

Blackburn, R. (1970, January 16). 'MacIntyre, the Game Is Up', Black Dwarf, 11.

Blackledge, P. (2005). "Freedom, Desire, and Revolution: Alasdair MacIntyre's Early Marxist Ethics," *History of Political Thought, 26*(4): 696–720.

Blackledge, P., & Davidson, N. (2008a). *Alasdair MacIntyre's Engagement with Marxism: Selected Writings 1953–1874*. London: Brill.

Blackledge, P., & Davidson, N. (2008b). "Introduction." In P. Blackledge & N. Davidson (eds.), *Alasdair MacIntyre's Engagement with Marxism: Selected Writings 1953–1874*. London: Brill.

Brown, A. (2017, 27 May). "Leaked documents reveal counterterrorism tactics used at standing rock to "defeat pipeline insurgencies." *The Intercept*. Retrieved from https://theintercept.com/2017/05/27/leaked-documents-reveal-security-firms-counterterrorism-tactics-at-standing-rock-to-defeat-pipeline-insurgencies/ 18 June 2019 11:14 AM EST.

Brown, A. (2018, January 9). "Five spills, six months in operation: Dakota access track record highlights unavoidable reality—Pipelines leak." *The Intercept*, Retrieved from https://theintercept.com/2018/01/09/dakota-access-pipeline-leak-energy-transfer-partners/, 18 June 2019 11:17 AM EST.

Burger et al., (2012). "The macroecology of sustainability." *PLOS Biology, 10*(6). Retrieved from https://journals.plos.org/plosbiology/article?id=10.1371/journal.pbio.1001345. Last retrieved 10 July 2019, 1:28 PM EST.

Business Wire. (2019, May 4). "Energy transfer partners and sunoco logistics partners respond to the statement from the department of the army." Retrieved from https://www.businesswire.com/news/home/20161204005090/en/Energy-Transfer-Partners-Sunoco-Logistics-Partners-Respond. Last Retrieved, 18 June 2019 11:13 AM EST.

Dalrymple, A. (2016, August 18). "Pipeline route plan first called for crossing north of Bismarck." *The Bismarck Tribune*. Archived from the original on November 1, 2016. https://web.archive.org/web/20161101185338/http:/bismarcktribune.com/news/state-and-regional/pipeline-route-plan-first-called-for-crossing-north-of-bismarck/article_64d053e4-8a1a-5198-a1dd-498d386c933c.html Last Retrieved, 18 June 2019 11:15 AM EST.

Federici, S. (2014). *Caliban and the Witch: Women, the Body, and Primitive Accumulation.* New York: Autonomedia.

Frazer, E., & Lacey, N. (1993). *The Politics of Community: A Feminist Critique of the Liberal-Communitarian Debate.* Hemel Hempstead: HarvesterWheatsheaf.

Frazer, E., & Lacey, N. (1994). "MacIntyre, feminism and the concept of practice." In J. Horton & S. Mendus, (eds.), *After MacIntyre: Critical Perspectives on the Work of Alasdair MacIntyre.* Cambridge: PolityPress, pp. 265–282.

Hoffman, T. J. (1997, October). "Moving beyond dualism: A dialogue with Western European and American Indian views of spirituality, nature and science." *Social Science Journal, 34*(4): 447–461.

Holly, M. (1994). "The persons of nature versus the power pyramid: Locke, land, and American Indians." *International Studies in Philosophy, 26*(1): 13–31.

Horkheimer, M., & Adorno, T. (2002). *The Dialectic of Enlightenment: Philosophical Fragments,* Palo Alto, CA: Stanford University Press.

Ingram, D. (1995). *Reason, History, and Politics: The Communitarian Grounds of Legitimation in the Modern Age.* Albany: State University of New York Press.

Knight, K. (2011). "Revolutionary Aristotelianism," In P. Blackledge and K. Knight (eds.), *Virtue and Politics: Alasdair MacIntyre's Revolutionary Aristotelianism,* Notre Dame: Notre Dame University Press, pp. 20–34.

Kymlicka, W. (2002). *Contemporary Political Philosophy.* Oxford: Oxford University Press.

Liu, L. (2016, September 13). "Thousands of protesters are gathering in North Dakota — and it could lead to 'nationwide reform'." *Business Insider.* Retrieved from https://www.businessinsider.com/photos-north-dakota-pipeline-protest-2016-9/#while-members-of-the-standing-rock-sioux-tribes-began-protesting-the-project-as-early-as-april-protests-heated-up-in-august-as-numbers-increased-to-the-thousands-. Last Retrieved, 18 June 2019 11:16 AM EST.

McMylor, P. (1994). *Alasdair MacIntyre: Critic of Modernity.* London: Routledge

MacIntyre, A. (1959). "Notes from the Moral Wilderness: 2." *New Reasoner, 8*: 89–98.

MacIntyre, A. (1960a). "Freedom and revolution." *Labour Review, 5*(1): 19–24.

MacIntyre, A. (1960b). "Breaking the chains of reason." In E. P. Thompson (ed.), *Out of Apathy,* London: Stevens and Sons, pp. 195–240.

MacIntyre, A. (1978). "How to identify ethical principles." In The National Commission for the Protection of Human Subjects of Biomedical and Behavioral Research (eds.) *The Belmont Report: Ethical Principles and Guidelines for the Protection of Human Subjects of Research.* DHEW Publication No. (OS) 78-0013, pp. 101–10-41.

MacIntyre, A. (1985). "Rights, practices, and marxism." *Analyse & Kritik, 7*: 234–248.

MacIntyre, A. (1988). *Whose Justice? Which Rationality?* Notre Dame, IN: University of Notre Dame Press.

MacIntyre, A. (1991). "An Interview with Alasdair MacIntyre," *Cogito, 5*(2): 67–73.

MacIntyre, A. (1994). 'Interview with Professor Alasdair MacIntyre.' *Kinesis, 20*: 34–47, 43.

MacIntyre, A. (1998). "Politics, philosophy, and the common good." In K. Kelvin (ed.), *The MacIntyre Reader,* Notre Dame, IN: University of Notre Dame Press, pp. 235–252.

MacIntyre, A. (1999a). *Dependent Rational Animals: Why Human Beings Need the Virtues* LaSalle, IN: Open Court Publishing.

MacIntyre, A. (1999b). "Social structures and their threat to moral agency." *Philosophy, 74*: 311–329.

MacIntyre, A. (2007). *After Virtue: A Study of Moral Theory.* Third edition. Notre Dame, IN: University of Notre Dame Press.

MacIntyre, A. (2011). "How Aristotelianism Can Become Revolutionary: Ethics, Resistance and Utopia." In P. Blackledge and K. Knight (eds.), *Virtue and Politics:*

Alasdair MacIntyre's Revolutionary Aristotelianism, Notre Dame: Notre Dame University Press, pp. 11–19.

MacIntyre, A. (2017). *Ethics in the Conflicts of Modernity: An Essay on Desire, Practical Reasoning, and Narrative.* Cambridge: Cambridge University Press.

Mufson, S., & Eilperin, J. (2017, 24 January). "Trump Seeks to Revive Dakota Access, Keystone XL Oil Pipelines," *The Washington Post*, Retrieved from https://www.was hingtonpost.com/news/energy-environment/wp/2017/01/24/trump-gives-gree n-light-to-dakota-access-keystone-xl-oil-pipelines/?utm_term=.203bc818f984. Last Retrieved, 18 June 2019 11:18 AM EST.

Mulhall, S., & Swift, A. (1996). *Liberals and Communitarians* 2nd ed., Oxford: Blackwell.

Naureckas, J. (2017). "Hiding DAPL violence behind 'nothing to see here' headlines." *Extra, 30*(1).

Nicholas, J. (2012). *Reason, Tradition, and the Good: MacIntyre's Tradition-Constituted Reason and Frankfurt School Critical Theory.* Notre Dame, IN: University of Notre Dame Press.

Nicholas, J. (2014). "Toward a radical integral humanism: MacIntyre's continuing marxism." *Studia Philosophica Wratislaviensia*, Supplementary English edition on-line.

Nicholas, J. (2015). "The common good, rights, and catholic social doctrine: Prolegomena to any future account of the common good." *Solidarity: The Journal of Catholic Social Thought and Secular Ethics, 5*(1): Art. 4.

Nicholas, J. (2017). "Refusing polemics: Retrieving Marcuse for MacIntyrean Praxis." *Radical Philosophy Review: Special Issue: 50th Anniversary of Marcuse's One Dimensional Man, 20*(1): 185–213.

Nussbaum, M. (1989). "Recoiling from reason." *The New York Review of Books, 36*(19): 36–41.

Pope, F. (2015). *Laudato Si: On Care for Our Common Home.* Huntington, IN: Our Sunday Visitor.

9

NATURE, LIBERTY, AND ONTOLOGY

Why nature experience still exists and matters in the Anthropocene

Piers H.G. Stephens

Amongst environmental thinkers and activists, it is commonplace to assert that time spent engaged with non-human nature is rejuvenating to the capacity to perceive novelty, to gain creativity, or to place one's social anxieties in a proper context. Though environmentalism is generally conceived of as a modern movement, it is worth noting that these intuitive connections between nature and liberty have a much longer history, even reaching back into antiquity. In this chapter, I trace the development of that history, beginning with two pre-modern traditions of thought in which direct experience of nature was associated with the stimulation, construction, and defence of human liberty – agrarian thought and sylvan liberty – and show how they offer different concepts of human freedom and flourishing. Next, I link these traditions to Bryan Norton's work on sustainability, with its emphasis on options for human developmental growth explaining how the transformative possibilities of nature experience may help defend the non-human world. This connects the continued existence and experience of non-human nature to a conception of human liberty, linked to human flourishing. Third, I argue that space exists for an ontology of nature that avoids the excesses of man/nature dualisms on the one side, and pure naturalism or social constructionism on the other. In such an ontology, naturalness is present in *relative* terms on a tripartite spectrum: naturalness is present to the extent that an item or area has not been transformed in accordance with certain historically specific, objectifying, and anti-naturalistic types of human instrumental rationality. This shows, I argue in conclusion, the worth of at least some ontological conceptions of non-human nature in the Anthropocene, and reaffirms linkage between nature experience and human liberty.

Nature and liberty: Two traditions

The relationship between non-human nature and freedom has a complex history. An initial clue to the importance of the concept of human liberty in environmental thought is the presence of anarchist thought running throughout today's green movement and its historical antecedents. It is present in Kirkpatrick Sale's bioregionalism and Murray Bookchin's social ecology, its long pedigree runs back through Kropotkin and Thoreau to, if Peter Marshall's literary interpretation is accepted, William Blake (Marshall, 1994), and its philosophical heritage arguably extends back further still. Indeed, I hold that the real point of linkage between green thought and anarchism lies not in the political repudiation of the state, but with the conception of an unfolding, dynamic, and harmonious (because spontaneous) order of nature. Andrew Belsey notes that pre-Socratic nature, *kosmos*, embodies notions of processively unfolding order, morality, harmony, and beauty immanent in the natural world (Belsey, 1994, pp157–60). This linkage of *kosmos* with the anarchist ideal of natural, spontaneous, and organic human development, Belsey argues, still survives as a thread of resistance in anti-authoritarian writings throughout the Western philosophical tradition since that time, though its associations with natural vitality are not always cosy or egalitarian. Speaking mainly of inner rather than outer nature, Charles Frankel has expressed the essence of this tradition as being the idea of "nature as a principle of development," which tells us that "there is something under the human skin with its own vitality, something not wholly malleable, not susceptible to Skinnerian conditioning except at the price of destroying spontaneity, talent, zest, vitality"; the notion of nature here vitally celebrates the natural as a sphere which outruns the power of human rule and artifice (Frankel, 1976, p104). This conception is associated with a preference for the virtues and understanding induced by active rural life rather than the manipulative corruption of the city. It is here that we may differentiate two traditions of thought that associate nature as a touchstone of liberty: the agrarian tradition and the sylvan liberty tradition. In different ways, both share a suspicion of the vices and artificialities of the city, and it is to the marks of the two traditions that we now turn.

The agrarian tradition

The agrarian tradition can arguably be traced back into classical Greece, with Victor Davis Hanson regarding the family farm as the hitherto under-recognized origin point for the rise of the *polis*. Hanson observes the way in which the novel Greek agriculture of trees and vines meant that, of necessity, farmers had to invest labour and capital in a particular place on a lifelong basis, a shift that meant "changing his way of thinking from mere production to stewardship of a lifetime's investment" and thus the development of a set of accompanying virtues. Indeed, this was the necessary backdrop to the more familiar flowering of Greek intellectual culture, for the new agrarianism "created the surplus, capital, and leisure that lay behind the entire Greek cultural renaissance" while being "highly flexible and decentralized economically, socially egalitarian, and politically keen to avoid the accumulation of power by a nonagricultural elite." It should thus be unsurprising that "the later *polis* Greeks envisioned the rise of agrarianism – which had created their city-state – primarily in moral terms" (Hanson, 1995, p43, 45). Certainly, the resultant agrarian notion of the farm as a moral counterpoint to the vain ambitions of city life was one that then manifested itself in Stoic thought much longer, running throughout the Roman period. Consider the following lines:

> If you can face the prospect of no more public games
> Purchase a freehold house in the country. What it will cost you
> Is no more than you pay in annual rent for some shabby
> And ill-lit garret here. A garden plot's thrown in
> With the house itself, and a well with a shallow basin –
> No rope and bucket work when your seedlings need some
> Water!
> Learn to enjoy hoeing, work and plant your allotment
> Till a hundred vegetarians could feast off its produce.
> It's quite an achievement, even out in the backwoods,
> To have made yourself master of – well, say one lizard, even
> *(Juvenal, 1974, p95)*

These lines, which might pass for verse from a 1970s self-sufficiency enthusiast, were written by the Roman poet Juvenal in his Third Satire, *Against the Life of the City*, in approximately 110 AD. They validate an explicitly *agrarian* life: the counterpoint to the excesses of city life that is being used is not a dichotomous one between nature/human, nature/culture, or nature/artefact, as commonly encountered in the American wilderness tradition. Rather, the theme is better understood if we conceive of the concept of nature as being ontologically defined along a spectrum rather than dualistically, with its manifestations running from the overwhelmingly humanized city through to the respectfully stewarded farm and then on out to the wild areas, to which we shall soon turn.

Whilst the agrarian tradition is familiar, its direct relevance to contemporary environmental concerns might be questioned given the steep ongoing decline in the number of the populace engaged in farming in the affluent world. However, as the intermediate example of Juvenal's writing suggests, many similar features of association might be drawn from relevantly similar, but smaller scale and increasingly popular activities, such as gardening. I have argued elsewhere (Stephens, 2016a, pp183–197) that the activity of gardening shares many of the liberating features of traditional agriculture, and three of these features matter here.

First, gardening helps train greater humility and openness in perception. Robert Pogue Harrison maintains that "gardening brings about a transformation of perception," in which the eye observing nature's forms "looks into the depths from which they stake their claims on life and from which they grow into the realm of presence and appearance," a process that uniquely enables us "to realize what life is up against in its struggles to affirm its rights in the ground we tread upon" (Harrison, 2008, pp30–31), teaching both humility and optimism through the activity. Such learning may also intertwine with familiar broader arguments about nature and human wellbeing, such as E.O. Wilson's biophilia hypothesis and Louv's claims about "nature deficit disorder" (Wilson, 1984; Louv, 2006; Pretty, 2007), and suggests that such activities foster the idea of belonging in place at an unusually deep psychological level.

The second feature is that agricultural and gardening practices involve depth of experience. Stephen Kaplan has drawn on William James's work about attention to argue for what he calls his Attention Restoration Theory (ART), which maintains that prolonged use of directed attention – the type of willed attention associated with many contemporary technological environments – results in mental fatigue, as well as reduced concentration and irritability, but that activity in coherently different and humanly compatible natural environments are therapeutically restorative, relieving mental fatigue by allowing directed attention to rest, and replacing it with spontaneous fascination and engagement (Kaplan, 1995). Supporting the idea of a primordial psychological link, Harrison notes how garden imagery is pervasive in mythology, religion, and allegory, remarking especially upon the parable Gaius Julius Hyginus told of the goddess Cura, "Care," who shaped wet clay into a human form before Jupiter gave it life by adding his spirit. Cura, Jupiter, and Earth all then disputed over whose name should be given to the creature, finally submitting to the judgment of Saturn that the human creature should at death finally return in body to Earth, from whence s/he began, whilst the spirit should return at death to Jupiter, but throughout life s/he should be possessed by Care, the first shaper, and be called *homo* after the *humus* of the Earth (Harrison, 2008, pp40, 5–6). Thus, it would seem, the need to work with care upon the Earth emerges in mythology as the first and essential defining task of humanity.

This leads us to the third feature, a point made by Shane Ralston about gardens but which might also be said of smaller traditional farms, which is their existence as a "moral space," which can provide the "material and intellectual conditions for an entire community to flourish" and represent "how humans cultivate their own potential as moral agents, taking into consideration the interests of others" (Ralston, 2013, p77). The tradition of agrarianism and its gardening affiliate helpfully encourages us, as Roger King observes, to morally "think about the relationships between the kinds of spaces we occupy on a daily basis and the wild spaces that environmental ethics most wants to protect" (King, 2003, p4).

What Juvenal and his Stoic contemporaries had in mind, then, was a notion that is still practical in the modern era: a view that regular, respectful work within the rhythms of the natural world has a regulative effect on the human psyche, channelling natural spontaneity and zest into positive virtues of conduct. Indeed, as several thinkers have suggested (Ball, 2000; Cannavo, 2010; Thompson, 2014), in an American context, we see a later emanation of these ideas in the Jeffersonian tradition, in which several connections link agrarian nature and liberty: first, nature is represented as a guiding force that the agriculturalist develops the non-capricious virtues through attending to; second, such attendance may be encouraged and passed down through generations through suitable ownership patterns; third, these virtues are appropriate for a democratic republic; and fourth, agrarians are regarded as free and independent through not being dependent on others for direct sustenance. Nature experience thus links to liberty in the agrarian tradition by stimulating a virtuous and flourishing political life, tied to the rhythms of the natural world in a non-dualistic relational manner.

The sylvan liberty tradition

We now turn to the tradition of sylvan liberty. This is more associated with the transformative powers of truly wild nature, differing from agricultural land by occupying an ontological place amongst the less or least humanly transformed of landscapes, paradigmatically represented by the forest. The first point to note here is the cross-cultural pervasiveness of the view that woodland is primordial in human origins and thus in some sense both an origin and an antithesis to the life of the developed city. Norse mythology traced the beginnings of the world to the cosmic ash tree Yggdrasil, whilst for the Greeks, Artemis and Dionysus were deities of the grove and hunting and "represented in turn the enigmatic ideal and the primal urges that come from the depths of the inner self, mirrored in the depths of the forests" (Hayman, 2003, p3).

Moreover, before Aristotle's foundational attempt to give form a logical priority over substance in the redefinition of phenomena, classical Greece had seen arguments like those of the proto-materialist Antiphon, who argued that substance – *physis* – is more important and enduring than form. Antiphon observed that if a buried bed of rotting wood threw out a shoot, that which would survive would be the substance, wood, and not the form, bedstead, but his claim was refuted by Aristotle's argument that when we speak of the nature of something, we must speak of its formal properties, which are transmitted by the *telos* of form, rather than by its substance. Substance as such, or in modern parlance "matter," is something about which we cannot speak in any logical manner, maintains Aristotle, but as Robert Pogue Harrison observes, "there is one word that Aristotle could not avoid using when he spoke about the unspeakable – *hyle*." Aristotle is "the first to give the word its philosophical meaning of 'matter'" and yet "*hyle* in Greek does not originally mean matter, it means forest" (Harrison, 1992, p28).

The originary foundation of matter, of all that *is*, of that from which we are made and to which we return, *hyle* actually refers to a forest! Ancient Rome too regarded itself as born from the wild forest through the myth of Romulus and Remus, suckled by a she-wolf in the forests of Latium, and the Latin perspective also had some universalist implications. The poet Lucretius initiated a long-running set of associations of woodland life in his claims that early humans, and by implication more primitive contemporaries too,

> huddled in groves, and mountain-caves, and woods, and 'mongst the thickets hid their squalid backs, when driven to flee the lashings of the winds and the big rains. Nor could they then regard the general good, nor did they know to use in common any customs, any laws: whatever of booty fortune unto each had proffered, each alone would bear away.
>
> *(Lucretius, 2018, Book V, lines 954–961)*

With these associations widely in play, we should not be too surprised that deforestation was seen by the Romans as a significant civilizing element in their imperial campaigns, an association of forest conquest with civilization that remains in many minds even today; however, even desert peoples associated trees with the psychology of human origins, as with the Genesis story of Adam and Eve.

Outlaw freedom

These associations of the forest with untamed nature beyond human law set up a further type of nature–liberty connection, one most fruitful for us to explore in the period immediately prior to the large-scale English enclosure of the commons which birthed capitalism. The 17th century was a heyday for the romanticized figure of the outlaw, totemically represented by such figures as Robin Hood and William Wallace, and the popular ballads of sylvan liberty that were composed about them and their kind. Christopher Hill observes that the "forest was always seen as an area of imagined freedom adjacent to a town" (Hill, 1996, p72), and although forests had served as ground for tribal attacks on occupying forces during the Roman period, medieval England of the 14th and 15th centuries saw the birth of the association between forests and rebellious outlaws, who were often lionized by the common people. Eric Hobsbawm refers to "social bandits" in this context, the "peasant outlaws whom the law and the state regard as criminal but who … are considered by their people as heroes, as champions, avengers, fighters for justice, perhaps even leaders of liberation" (Hobsbawm, 1969, pp17–19). In similar vein, Richard Hayman observes that several 15th century popular ballads portrayed Scottish hero William Wallace as waging a forest-based guerrilla war against the English, and that it is invariably "in the woodland haunts that Wallace is able to rally men behind him, emphasizing the association of wild nature with natural justice." Moreover, such tales were widespread and extremely popular: "all outlaw ballads and tales had some suitable oppressor to make their cause a just one" (Hayman, 2003, p55).

Turning to the most famous figure of sylvan ballads, Robin Hood, we do *not* see him originally redistributing wealth, nor is he characterized as a nobleman in any of the early stories, but he is unquestionably portrayed as a symbol of self-reliant independence and resistance to corrupt authority, including clerical authority. (It is striking that Robin Hood ballads were associated with the May Day games in which authority was regularly mocked, and that at a time when religious division formed the primary national tension, these games were suppressed under both Catholic and Protestant rulers.) Most tellingly, Hill observes that the Robin Hood ballads were especially popular in the 16th and early 17th centuries – exactly parallel with the rise of the enclosure movement, and the increase in urban immigrants seeking work after losing their access to land. Moreover, the outlaws' colours of Lincoln green were also those of the anti-enclosure Leveller movement, though this may have been coincidental (Hill, 1996, p78, 75).

At the time of early modernity, the forest thus represented primordial natural vitality, liberty, and resistance to conventional law in the popular mind, as well as having strong implications of natural justice, and these associations were common to more conservative writers as well. Shakespeare's Forest of Arden in *As You Like It*, for instance, is a place in which conventional roles and identities of nobility slip; similar distortions occur under magical influence in the forest scenes of *A Midsummer Night's Dream*, whilst the usurping tyrant *Macbeth* is overthrown once "Birnam Forest come to Dunsinane" (Shakespeare, 1970, p1024) via Macduff's army, though in these cases the crises' resolutions operate to ultimately restore the ideal conventional order. The association of natural law with the raw nature of the forests, and with transformations that counterpoint the political tensions of the city, is in fact a central idea in the period leading up to the triumph of the new merchant class and of enclosure in the latter 17th century, and had constituted a core and long-standing cultural connection between ideas of raw nature and of human original liberty in Western political thought for centuries prior to that time, but one that dropped out of mainstream use thereafter. Hayman summarizes this linkage in relation to the figure of the forest outlaw:

> The moral of all outlaw ballads is that the good win and evil is defeated. They celebrate the freedom of the individual against corrupt institutions. The individual prepared to stand up for justice will attract a loyal following or, to set it in its conservative guise, simple folk want moral leadership and are instinctively drawn to real against false leaders. They believe in social justice but do not trust the agents of its institutions to deliver it. Outlaw ballads therefore say as much about contemporary attitudes to the law as they do about greenwood idylls. The independent spirit of the outlaws is characterized by their existence in the greenwood, where the ability to endure real hardships is a badge of integrity. Greenwood men are not bound to one place like other men; their peripatetic lives are freer than any other way of life, and they enjoy the savor of unpredictability. Perhaps they convince themselves that they live freely as men must have lived before they tilled the soil. In their idealized form outlaws do not have property

to emphasize their status and as barriers against congress with lower ranks, nor do they dress differently to mark their status out to strangers. Unlike other men, they live by their own efforts, share the fruits of their labor in a spirit of comradeship, and, like huntsmen, cannot be accused of the sins of idleness or luxury. These qualities become more accentuated when they are contrasted with officers of the law and high ranking churchmen, who abuse their privileges and treat justice as a commodity.

(Hayman, 2003, pp56–57)

From freedom in nature to freedom as transforming nature

This summary indicates that two interlinked but distinguishable traditions of thought, agrarianism and sylvan liberty, fundamentally connected the ideas of external nature and of liberty, and that the origins of these can be identified from an early stage in Western thought.

Yet aside from certain maverick figures like Rousseau, Jefferson, and Thoreau, they largely faded out of mainstream political philosophy after the late 17th century, and, in my view, this fading stemmed from the intertwined triumphs of Newtonianism in science and Lockean philosophy in politics. Though space precludes long explanation, I have argued elsewhere (Stephens, 1999, 2001) that Locke's political philosophy is the most significant intellectual contribution from that period for motivating and justifying an agenda of increasing economic productivity and the humanizing transformation of nature through labour, an agenda in which the transformation of nature for economic gain was to become the paradigmatic example of manifesting human liberty. Locke clearly rejected the sylvan liberty tradition, and openly expressed his disdain for what he called the "irrational untaught Inhabitants" of "Woods and Forests" who "keep right by following Nature" in the *First Treatise* (Locke, 1988, p183). In this he expressed a view common among the affluent that forests and wastes encouraged idleness, irrationality, and immorality, and moreover that enclosure of land actually benefited the occupants in the longer term through increased production and a trickle-down effect. Moreover, not only did his philosophy embody a Baconian emphasis on transforming and commanding non-human nature, but in doing so it manifested strongly dualistic assumptions about the relationship of man and nature, and about the proper priorities between types of liberty in what was to become the liberal democratic tradition. Locke's relegation of non-human nature as an inspiration to liberty, and his invocation of nature-transforming labour as instead being the core of value and the optimal manifestation of rational freedom, fitted with the rise of atomism and mechanism, and their grand unification in the Newtonian vision of the world. The Lockean/Newtonian vision is one in which the implications of spontaneity and vibrancy found in the *kosmos* conception of nature are utterly removed. Freya Mathews appositely describes this perspective and its implications for our moral view of nature:

In pre-scientific thought, Nature had been richly informed with *telos*, and with principles of spirit and agency. Human beings existed in an intricate web of spiritual and teleological relations with the natural world. From the mechanistic point of view however, Nature consists of matter, and matter is insensate, dead, drab, unvarying, devoid of interests and purposes. This draining off of spirit from matter was naturally expressed in mind-matter dualism: the human mind had to become the repository of spirit once Nature had become the arena of blind matter in motion. Dualism gave expression to the mechanistic idea that matter was essentially utterly unlike ourselves: we are essentially identified with spirit, and matter was conceived as in every respect antithetical to spirit. As such – as the insensate, brute and blind, the inert and formless, the non-self, the Other, the External – matter of course ceased to be an object of moral concern or interest.

(Mathews, 1991, pp31–32)

The important aspects of human freedom now take wing not from an interaction with nature that gives wisdom, nor from a location within sylvan nature that is outside the realm of arbitrary human law, but instead in the human mind and its capacity to transform nature's brute matter into items of our own devising. In an inert determinate world in which the human soul alone manifests freedom, economic activity that transforms nature in accordance with human will *becomes* the paradigmatic liberty, the freedom that defines us as uniquely human and thus, in an experientially self-reinforcing process, further confirms the reality of our uniquely human freedom. And this dynamic, encouraging endless transformation of nature and increased economic production, is identifiably a key aspect of the culture of contemporary capitalist politics and society, one that stands in a relationship to sustainability and the limits of our planetary capacities that is at best tense, at worst hugely destructive.

Sustainability and the preservation of liberty

Fortunately, there are alternatives to this dualistic ontology and more eco-friendly conceptions of liberty and human fulfilment are still available to us. Bryan Norton's work on sustainability suggests one way in which we can make use of such conceptions.

Norton has outlined two basic defining options as available: weak sustainability, defined by intergenerational utility comparison over time (UC), and strong sustainability, his variant of which is defined by listing stuff (LS) that should be saved for future generations. The "stuff" in the latter consists of natural items whose loss would "result in the diminution of the quality of future lives, *regardless of the amount of compensation/wealth that is provided as a substitute for the lost features*." Norton argues that we may be obliged "not to unduly narrow the range of options and opportunities open to future people," (Norton, 1999, pp130–131) and makes the analogy with a rich sexist widower who bequeaths his daughters an excellent income for life on condition that they not seek an education: though they are financially affluent, harm is done through narrowing the range of a person's developmental opportunities.

Norton's suggestion resonates with liberal ideals of development, with its view that "what is valua*ble* for human enrichment, has to be expanded ... in terms of a greatly extended notion of human interest" (Rosenthal & Buchholz, 1996, p43), for he argues that strong sustainability should specify certain features of the natural world, associate them with measurable indicators and protect them as representing options and opportunities that future people would be worse off without. Drawing on this idea that natural areas represent human developmental opportunities, I maintain that nature is significant in enabling and expanding human liberty. The answer to the question "why should those who value liberty protect nature?" is thus "because they protect developmental liberties by doing so."

Nature on a spectrum of instrumental rationality

Earlier work of mine argued that nature should be defined across a tripartite ontological spectrum in which naturalness is defined by *the extent to which an item or area is untransformed* and *not instrumentalized according to the dictates of particular historically specific types of anti-naturalistic instrumental rationality*, with the spectrum itself derived from William James' model of the human cognitive faculties in their perceptual operations of truth-seeking (Stephens, 2000). It is the processes of instrumentalization and artefacticity that are thus the antitheses of the natural; the mere presence of human input, such as a ship in the ocean deeps, does not eliminate ontological naturalness unless we wrongly assume an absolutistic dualism in which *any* human input is seen as utterly transformative. In this model, rather, the only nature we can have access to is in a certain sense humanized through the ways in which our consciousness has learned to select according to our past human experiences and ongoing needs, but nonetheless some experiences are more raw, less "cooked" by past instrumental experience, expectation and planning than others, and similarly we may say that certain areas of nature are less instrumentalized and artefactualized than others. With this as a base point, we can lay out a spectrum in which we have *relatively* untransformed nature at one pole, borderline places such as traditional farms and country paths, as well as objects that are made with a spirit of respect towards nature acting as intermediate points on the spectrum, and finally the world of completed artefacts and radically instrumentally transformed goods at the extreme of artefacticity, the realm I call the artificial, at the other pole.

This framework is important in that it clarifies issues in relation to the postulated framework of the Anthropocene. For the morally transformative impact often attributed to the Anthropocene is usually based around the idea of what may, following Bill McKibben's phrase, be called the "end of nature." On this "end of nature" conception, there is no more wild nature because human impact is everywhere, and in the absence of anything without human influence, there is no such thing as any independent nature to speak of coherently, and ideas of nature experience or wilderness should be jettisoned as anachronistic. But adoption of the ontological scheme that I have suggested allows an easy response, because it already notes the *relative* character of nature's definition and presence. The unspoken premise in the idea that the Anthropocene undermines coherent talk of nature is the idea that human presence is a sort of impurity that erases the non-human, and the notion of (im)purity running here is dualistic: rather than being the sort of *relative* purity associated with environmental evaluations, such as "this water is relatively pure of mercury contamination," it tacitly models purity in an either/or fashion, like virginity, so that the ontological status of nature is wholly changed merely by the smallest touch of human influence. But this is to make enormous question begging assumptions. For to say that there is no part of the earth that has not felt human influence is not remotely to show that non-human nature no longer has elements of independence in its dynamics. Indeed, as Simon Hailwood has observed, not all meanings of the diffuse term "nature" have the implications of necessarily untouched purity, and in fact in evaluating evidence about climate change dynamics and other aspects of the postulated Anthropocene era, "there are questions about the extent of the change and the extent to which the change is anthropogenic" (Hailwood, 2016, p51). Yet these questions would not even make sense if there were *not* continuing independent dynamics in the natural world with which human impacts are interacting. The meaningful point is thus not the presence or absence of *some* human influence per se. Rather, it is the character of that influence: that is, the extent to which it is transformative and instrumentalizing. If we employ the tripartite spectrum scheme I have described then a more defensible account of nature's ontology results, whilst the meaningfulness of nature experience and its connection to liberty may be retained.

The relative absence of these instrumentalizing human elements, I hold, is critical not only for ontologically defining "nature," but for indicating what is vital in nature experiences: the extent to which they grant possibilities of new perception through non-instrumentalized immediate experience. As known since William James's pragmatism, consciousness is selective and instrumental, excluding extraneous information, reinforcing attention to already existing interests, goals and associations, and building up habitual patterns. These patterns of awareness guide us in our customary orientations, but they are necessarily instrumental and drawn from past analogies and experiences. They are thus prima facie committed to our existing framework of ends, functioning to streamline our awareness and edit out novelty as irrelevant unless it impacts existing goals and interests. What is reduced in such instrumentally dominated awareness, and reduced ever more strongly as habitual thinking grows over a lifetime, is *the capacity to step outside the existing framework of priorities*, to question the existing instrumental framework and view issues afresh, to gain openness, to perceive novelties that might make us revise our views. We thus build a shell of habitual perceptions and pre-set means-end reasoning, but at the cost of sensory vividness, flexibility, the capacity to learn afresh, and some restriction of the imagination.

Pragmatic naturalism

A pragmatic naturalist antidote to this limiting tendency exists, however, as I have argued previously (Stephens, 2009), a way whereby we may continue to protect and enlarge our freedom. It lies in types of experiences in which we are more receptive to direct sensation with minimal overlay from ready-made descriptions and pre-set instrumental purposes: quiet attention to snowflakes falling outside, the sudden glimpse of a kestrel above, calm perception of a spider's web-work bridging two shrubs. These experiences draw us out of habitual instrumentalizing consciousness because they are *radically* noninstrumental, being neither means nor ends. They require from us a certain willingness to receive attentively without rushing to judgment and to engage reflectively with the impression thereafter, thus demanding a certain posture from the agent, but availability of such possibilities for growth also has a significant external component. For they occur in one context especially often: nature experience.

Continuing in Jamesian fashion, I further invoke a liberty argument I have suggested elsewhere (Stephens, 2016b). James found value in nature experience of the type described, noting "the intense interest that life can assume when brought down to the ... level of pure sensorial experience" and that such interest "starts upon us often from non-human natural things" (James, 1929, pp18–19, 9). I thus suggest there is a resonance between our primal inner nature – relatively unverbalised direct sensation – and the outer nature we encounter, a resonance that gives novelty and freshness to experience, allowing new perspectives to be developed. Supporting argument here may come from certain creativity tests. In these, a person is shown artefacts and asked to think of as many possible uses for them beyond their design function, with highest creativity given to the largest number of alternative uses. This is a test precisely because artefacts contain embedded purposes that immediately spring to mind, and they thus encourage thinking instrumentally but *against* originality of application. The test's point is that constant association of items with pre-set goals makes it hard to envision the items otherwise; taking this logic to its conclusion, an individual surrounded wholly by instrumental reasoning and its products comes to lack the capacity to see outside such a framework, to engage with spontaneous experience and learn. Something like this is already dangerously manifest in internet epistemic closure, whereby people become both enframed by screen technology and unwilling to look outside a small range of goals, sites and values as they need not be exposed to novelty or disagreement. I therefore suggest that nature experience contrasts to this because it lacks such pre-set human ends. The paradoxical "usefulness" of nature thus lies partly in its lying outside predefined use categories: its capacity to evoke open possibilities, to stimulate spontaneity, wonder, and non-instrumental relationship.

Conclusion

The traditions of agrarianism and sylvan liberty both celebrate these features of nature in pre-modern ways. The idea that nature experience counters internalised instrumentalism and thus psychologically assists freedom is abundantly affirmed in these traditions, which counterpose natural vitality to urban artifice and draw upon the classical notion of *kosmos* as a spontaneous, unfolding, and beautifully harmonious process. We can recover the critique of the manipulativeness of city life and the virtues associated with agrarian life, from antiquity to Wendell Berry. And we can recover the tales of sylvan liberty associating forests with freedom, wildness, and corrective transformation. I thus suggest that the historic persistence of such associations indicates their deep significance. The repudiation of instrumentalism, politicking and fashion in favour of nature's spontaneous novelty and authenticity suggests a further move, namely, highlighting imagination as stimulated by nature experience as a transformative agent in actualizing liberty's value. For imagination can operate not merely to envision alternatives but also to concretize freedom by doing so – for instance, shifting an agent from possessing negative freedom, an absence of external coercive blockage, to manifesting positive freedom, actualizing what was previously only a potentiality. Imagination thus connects to practical liberty, rendering abstract freedom into concreteness, and is psychologically fostered and encouraged by the continued presence and experience of maximally natural environments. A natureless world, like the life lived by Norton's uneducated affluent heiresses, would be one in which we had been deprived of vital options for transformative growth. If this argument is correct, then the defence of non-human nature and nature experience is itself partly the defence of human liberty and the possibilities of the human imagination.

References

Ball, T. (2000). "'The earth belongs to the living': Thomas Jefferson and the problem of intergenerational relations." *Environmental Politics*, 9(2): 61–77.

Belsey, A. (1994). "Chaos and order, environment and anarchy." In R. Attfield & A. Belsey (eds.), *Philosophy and the Natural Environment*. Cambridge: Cambridge University Press.

Cannavò, P. (2010). "To the thousandth generation: Timelessness, jeffersonian republicanism and environmentalism." *Environmental Politics*, 15(3): 356–373.

Frankel, C. (1976). "The rights of nature." In L. H. Tribe, C. S. Shelling, & J. Voss (eds.), *When Values Conflict*, (pp. 93–113), Cambridge, MA: Ballinger.

Hailwood, S. (2016). "Anthropocene: Delusion, celebration and concern." In P. Pattberg & F. Zelli (eds.), *Environmental Politics and Governance in the Anthropocene*, London: Routledge.

Hanson, V. D. (1995). *The Other Greeks: The Family Farm and the Agrarian Roots of Western Civilization*. Oakland: University of California Press.

Harrison, R. P. (1992). *Forests: The Shadow of Civilization*. Chicago: University of Chicago Press.

Harrison, R. P. (2008). *Gardens: An Essay on the Human Condition*. Chicago: Chicago University Press.

Hayman, R. (2003). *Trees: Woodlands and Western Civilization*. London: Hambledon & London.

Hill, C. (1996). *Liberty against the Law: Some 17th Century Controversies*. London: Allen Lane.

Hobsbawm, E. (1969). *Bandits*. New York: Dell.

James, W. (1929). "On a Certain Blindness in Human Beings." In *Selected Papers in Philosophy*, London: Everyman.

Juvenal. (1974). *The Sixteen Satires*. (Trans. Peter Green), Harmondsworth: Penguin.

Kaplan, S. (1995). "The restorative benefits of nature: Towards an integrative framework." *Journal of Environmental Psychology*, *15*: 169–182.

King, R. J. H. (2003). "Toward an ethics of the domesticated environment." *Philosophy and Geography*, *6*(1): 3–14.

Locke, J. (1988). *Two Treatises of Government*, (Ed. P. Laslett). Cambridge: Cambridge University Press.

Louv, R. (2006). *Last Child in the Woods: Saving Our Children from Nature-Deficit Disorder*. Chapel Hill, NC: Algonquin Books.

Lucretius. (2018, August 25). *On the Nature of Things*. (Trans. W.E. Leonard). Retrieved from http://classics.mit.edu/Carus/nature_things.html Book V, 2000.

Marshall, P. (1994). *William Blake: Visionary Anarchist*. London: Freedom Press.

Mathews, F. (1991). *The Ecological Self*. London: Routledge.

Norton, B. G. (1999). "Ecology and opportunity: Intergenerational equity and sustainable options." In A. Dobson (ed.), *Fairness and Futurity: Essays on Environmental Sustainability and Social Justice*, Oxford: Oxford University Press.

Pretty, J. (2007). *The Earth Only Endures: On Reconnecting with Nature and Our Place in It*. London: Earthscan.

Ralston, S. J. (2013). *Pragmatic Environmentalism: Towards a Rhetoric of Eco-Justice*. Leicester, UK: Troubador Publishing.

Rosenthal, S. B., & Buchholz, R. A. (1996). "How pragmatism *is* an environmental ethic." In A. Light & E. Katz (eds.), *Environmental Pragmatism*, (pp. 38–49), London: Routledge.

Shakespeare, W. (1970). "Macbeth." In P. Alexander (ed.), *William Shakespeare: The Complete Works*, London: Collins.

Stephens, P. H. G. (1999). "Picking at the locke of economic reductionism." In N. Ben Fairweather, S. Elworthy, P. H. G. Stephens, & M. Stroh (eds.), *Environmental Futures*, (pp. 3–23), London, UK: Macmillan.

Stephens, P. H. G. (2000). "Nature, purity, ontology." *Environmental Values*, *9*(3): 267–294.

Stephens, P. H. G. (2001). "Green liberalisms: Nature, agency and the good." *Environmental Politics*, *10*(3): 1–22.

Stephens, P. H. G. (2009). "Toward a Jamesian environmental philosophy." *Environmental Ethics*, *31*(3): 227–244.

Stephens, P. H. G. (2016a). "The tragedy of the uncommon: Property, possession and belonging in community gardens." In J. M. Meyer & J. M. Kersten (eds.), *The Greening of Everyday Life: Challenging Practices, Imagining Possibilities*, (pp. 183–197), Oxford: Oxford University Press.

Stephens, P. H. G. (2016b). "Environmental political theory and the liberal tradition." In T. Gabrielson, C. Hall, J. M. Meyer, & D. Schlosberg (eds.), *The Oxford Handbook of Environmental Political Theory*, (pp. 57–71), Oxford: Oxford University Press.

Thompson, P. B. (2014). "Thomas Jefferson's land ethics." In M. A. Holowchak (ed.), *Thomas Jefferson and Philosophy*, (pp. 61–78), Lanham, MD: Lexington Books.

Wilson, E. O. (1984). *Biophilia*. Cambridge, MA: Harvard University Press.

PART III
Resisting the undertow of modernity

10

LIBERATION FROM EXCESS

A post-growth economy case for freedom in the Anthropocene[1]

Rafael Ziegler

A variety of natural and social science assessments diagnose the unsustainability of the present: important environmental boundaries are already transgressed, minimum social requirements are not met for far too many, and inequality poses a risk to democracy. A study of the basic requirements of a good life, such as food, sanitation, and education, finds that of all 150 nations studied, no nation meets basic needs and does so in a way that is globally sustainable (O'Neill et al. 2018). It concludes that improvements in physical and social provisioning systems are needed if a good life for all is to be created and secured.

But perhaps this conclusion is too technical? Do we also need a shift in values? In the light of their growing ecological footprints, some identify Western, liberal societies as a main culprit, and put into question the value of liberty (and associated institutions of the "free" world). By contrast, others argue that liberty is a central value at the heart of plausible pathways towards post-growth and more sustainable ways of living. This chapter explores one variety of the latter perspective, presenting a freedom case for a post-growth plural system of market, public, and communal modes of provision.

The case is based on Paech (2012), and thus a prominent post-growth economy proposal emerging from a Western, industrialized nation state. "Facing the threat of drowning in a multiplicity of material options, she who reduces does not renounce but liberates herself from what is superfluous," writes Paech.[2] Paech's liberation case is based on a distinction of commercialized and de-commercialized areas. Closer examination of this distinction leads to a plurality of modes of provision and with it a freedom case for a post-growth economy, or so the next sections propose.

The case is compatible with a critical assessment of the history of freedom and its negative consequences for sustainability, which freedom sceptics point to. Economic freedom focused on extending markets and private ownership played an emancipatory role in overcoming feudalism and raising the standard of living. But this does not mean that such economic freedom did not have (unintended) consequences, or that it could not threaten individual sovereignty in the present (as Paech might put it). To wit, we are free to propose conceptions and consequences of freedom appropriate for our context – and this includes scrutiny whether what we call freedom is not unexamined luggage from the past.

Joan Robinson famously quipped: "Worse than being exploited under capitalism, is not to be exploited." Capitalist markets are so pervasive and important, that for better or worse, we seem stuck with them as a kind of lesser evil. While this interpretation is understandable in a context of emerging, industrializing nation states, we need to constantly reassess such interpretations. Worse than being exploited under capitalism, is to be dominated by the state? A "luggage" that seems to travel with some libertarians in disregard of the achievements of the welfare/social states emerging since the late 19th century? Worse than being exploited under capitalism, is to be subjected to the uncertainties created by an exit from the "safe operating space" of the Holocene? A "luggage" that seems to travel with top-down eco-state managers seeking to secure growth within limits.

Keynes (1930) proposes in the light of social anxiety and pessimism – depression then, unsustainability today – to "take wings into the future," consider long-term trends and possibilities, and the choices they reveal in the present. In this spirit, the next section turns to a vision of a post-growth economy that discovers in current economic and environmental crises the signs of freedom opportunity.

Liberation from Excess

In his bestselling book on post-growth economics, economist Niko Paech makes a liberation case for post-growth economics that has been both widely discussed and that is relatively "concrete" (Kallis et al. 2018)[3]. Paech's critique of the status quo in a nutshell (2012, p. 126f):

> Capitalist economies are based on a permanent increase in consumption goods. Except for a small group of wealthy people, most people can only buy these goods, if they earn sufficient income. This forces them into a dynamic of seeking more wage income resulting simultaneously in less time for non-wage labor and activities: raising children, cooking, care etc. Moreover, such activities are also increasingly marketed as services. Again consumers need additional money to pay for these services. In addition, the increase in consumer goods means that more time is required to choose among all the options, and to enjoy the numerous goods and services. But time is limited, and hence the increase in consumption possibilities does not correspond to an increase in real choice, utility from and enjoyment of goods. Rather, the result is an 'exhausted self' (Paech 2012, p. 128, quoting Alain Ehrenberg) that lacks control and sovereignty in matters

of consumption and production. There is an increasing consumption of anti-depressants, as well as medication of children with tranquilizing, performance-enhancing medication. However, this vicious cycle is not necessary: consumers can 'liberate themselves from affluence.'

(Paech 2012, p. 130)

Reduction of consumption to the right measure is the path to genuine enjoyment, or so he argues.

There is an intuitive appeal of this argument. Rather than lamenting consumer society, "we consumers" can take action ourselves with something as simple as dispensing with things not really needed, services not really wanted, and status questions that do not hold up to critical scrutiny. Incidentally, Keynes (1930), taking a long-term perspective, already noted that the love of money as a possession rather than as a means for something else, "will be recognised for what it is, a somewhat disgusting morbidity, one of those semicriminal, semipathological propensities which one hands over with a shudder to the specialists in mental disease." With their appeal to the "right measure," the writings of the economist Paech resonate with the Aristotelian vision of a good life, which discovers virtue not in excess but in the golden middle between the extremes.

However, is the appeal to individual sufficiency not heroic, and sociologically implausible in consumer democracies that structurally reproduce and require a system of growing and changing wants? Moreover, in the light of a plurality of lifestyles, and religious and philosophical views, who is to determine "right measure" and "good life"? What about people who disagree with Paech's version of the good life? What about happy consumers who neither feel exhausted nor lacking in sovereignty? Is the appeal to voluntary simplicity not sociologically and philosophically weak – too weak for the transformative goal of more sustainable ways of producing and consuming?

The appeal to voluntary simplicity is familiar from many de-growth and postgrowth proposals. It is typically complemented by considerations from political economy and critical reflection (Abraham 2015, p. 147f). Thus, it is noteworthy that Paech presents his argument as part of a structural post-growth society proposal. The remainder of this section will briefly outline this vision so as to prepare the liberation or freedom case for post-growth in the subsequent section.

Central to Paech's post-growth economy is a distinction of wage labour in money-based, exchange markets and labour in what he calls the de-commercialized area. People are to work in each for roughly 20 hours. The de-commercialized area is focused on subsistence labour, i.e. own production of food, manufactured things or services (such as education), the maintenance and repair of things, more communal use, as well as exchange of services in social networks and volunteering. This area makes people less dependent on wage labour. Provided with sufficient time, they have real alternatives for goods and services outside the market. Sufficiency becomes a structural opportunity:

de-materialization and getting rid of the "excess luggage" of affluent society, de-acceleration, and more time for relations, as well as a reduction of stimulation simply resulting from less exposure to money-based exchange systems, advertising, etc.

Wage labour in Paech's post-growth economy is part of a global division of labour that prioritizes regional economies and their value chains (supported by complementary, local currencies). The focus is on a circular economy with materially and culturally durable goods and services, on repair and modularity; on efficiency and consistent technologies, on conversion, renovation, re-building, and re-manufacturing. The global economy aims at an absolute decoupling of the economy from material growth, and a recovery of resources via a decrease in impervious surfaces/and reduction/rebuilding of artefactual structures, such as pavements (roads and parking sites) and industrial areas (airports). Thus, this sketch of Paech's post-growth economy indicates that liberation is not just a matter of individual consumer or lifestyle choices. Rather, choices depend on structure to become real options.

Liberation and post-growth economy: Philosophical underpinnings

So what are the real options? In response to the discussion of freedom and equality in political philosophy, Amartya Sen made a simple but basic point (Sen 1999): individual choices in production and consumption depend on personal, social and environmental traits that influence an individual's possibility of converting a good or service into desirable doings and beings. For example, if I am depressed, then even in the presence of material affluence my mental condition does not allow me to convert this affluence into desirable actions (and get out of the bed). Or, if I got out of bed and was passionate and knowledgeable about something – say experiments with dry toilets and fertilizer production for gardening – then my personal willingness and competence will be insufficient in the presence of norms banning or sanctioning the use of human faecal matter as fertilizer, etc. Formal and informal norms, just as psychological traits, play an important role in the transformation of formal options and resources into valuable functionings; or, as Sen put it, those doings and beings that persons have reason to value, their capabilities (Sen 1999).

But what are the doings and beings that persons have reason to value? The answer to this question is only limited by the imagination. Capability is therefore a wide category. However, there is a justice case for a subset of *central* capabilities. Martha Nussbaum (2006) has developed a basic theory of justice, proposing ten central capabilities for leading a life with dignity. The intuition behind the distinction of central and further capabilities is straightforward. Some choices and their associated doings and beings are central for agency and dignity, others are not. Say the capability of flying to 20 academic conferences per year. In turn, the question of effective conversion in relation to central capabilities is especially important. For example, do you have a right to free speech? And are the social norms such that you can exercise this right effectively?

The distinction of central and other capabilities introduces a sufficientarian threshold into the discussion. It resonates with the focus on subsistence production in Paech's work, and likewise with the much older discussion of "the eco nomic problem" as one about meeting the "absolute needs" of human beings, and not about those "relative in the sense that we feel them only if their satisfaction lifts us above, makes us feel superior to, our fellows" (Keynes 1930).[4]

However, the "athletic" (Cohen 1993) language of capabilities and agency requires a qualification. The members of a community lack agency, contingently or permanently, at least in some respects. Infants have limited capacities for deliberation and acting on reflected on goals. A severe accident can prevent a person permanently from deliberation and many actions. Moreover, legal practice recognizes further non-human animals. Environmental ethics has pushed the boundaries of community via sentientism to biocentrism to holism (Gorke 2010). Whatever one's positions in such debates, the community of justice is insufficiently captured by agency alone. We also need to consider contingent and permanent *patiency* (Ziegler and von Jacobi 2018). Partly out of mere self-interest – we are all patients to some extent – partly out of consideration for the ends of others. Nussbaum has recognized this point, and proposed a sentientist boundary of justice (Nussbaum 2006). However, if we are willing to consider patients, why not consider all beings whose "good" we can non-instrumentally investigate, to use Paul Taylor's well-known proposal? (Taylor 1986). A biocentric conception starts with a consideration of all living beings, and the non-instrumental consideration of what is good for them. Further exploration is beyond the scope of this paper,[5] but the basic point is clear: freedom and agency are part of a larger conception of well-being, including respect for the vulnerability of all members of the community.[6]

Focusing specifically on social norms, we can move from these points about liberty, agency and patiency to Paech's distinction of commercialized and de-commercialized work. "De-commercialized" is not a reference for a different sector or clearly defined area. Beyond saying what it is "not," specification is called for what this "other" area refers to. In fact, there is variety of norms, along with associated production and consumption choices, beyond the "commercialized" area of markets. This is a lesson from research at the intersection of economics and philosophy. It reveals a typology of modes of provision (Claassen 2009) that are very much present, albeit mostly implicitly, in Paech's proposal:

- *Self-provision.* The producer of the good consumes it herself. There is no exchange with others, no money involved.
- *Informal provision.* The exchange of goods is governed by social norms, for example in reciprocal gift-giving. No money is involved in exchanges, there is no expectation the respective giving is of the "same" value; reciprocity is "open to discretion as to value and time" (Offer 1997, p. 457).

- *Market provision.* Exchanges are based on price signals, and based on individual wants, preferences and purchasing power.
- *Public provision.* Resources are pooled (e.g. via taxes or mandatory labour) so as to provide goods to the respective group's members, be it public goods such as public radio or military defence, or the provision of specific goods linked to individual entitlements (e.g. public infrastructure for wheelchair accessibility).
- *Professional provision.* The good or service is based on the expert knowledge of the provider (e.g. doctors). Professionals are either paid directly by their clients or indirectly via public schemes (such as health insurance), private initiatives (such as philanthropic activities) or some combination thereof.

Drawing from the internal diversity and richness of the de-commercialized area beyond commercial market provision only, we can now turn to freedom arguments in support of a post-growth economy.[7]

Argument from diversity of subjects and their contexts

There is a huge diversity of individuals and their social and environmental contexts. According to the argument from diversity, at least in some contexts, more than one mode of provision is required to meet central capabilities. For example, care could be left to informal provision in the family and communal social networks (Claassen 2009, p. 436f). This might well be the best option for some individuals (both as recipients, as well as providers of care). However, some people might lack such a social context, or this context might be detrimental for their health (perhaps social relations are abusive or the care task is too complex, etc.). Accordingly, for somebody in this situation market or public provision might be important alternatives. Thus a plurality of modes in this example not only suggests better conversion possibilities given the diversity of individuals, it also provides choice for both recipients and providers.

In a rapidly changing world, no mode might be well-suited for the needs of some groups, say of refugees arriving in a new country. In such situations, more than one mode is prima facie less risky to secure the real freedom for individuals. An example from social innovation research further illustrates the point (Evers and Brandsen 2016, p. 165f): in urban social innovations across Europe, the innovators tend to relax the conditionality of admission into welfare programmes, including what counts as giving something back in return. The finding reflects the more discretionary understanding of reciprocity in informal provision discussed above. It thereby creates access to a service and good to people who are not in a position to meet the conditionality of the public welfare programme (perhaps only because the conditions are too demanding and they fail to show up with the right papers at the right time).

The argument from diversity calls for a consideration of what modes of provision help meet and secure central capabilities. As with central capabilities, such modes will be realized differently in a different context. However, the modes of provision offer a more comprehensive account of provision than a focus on markets alone. A priori, no mode is "better" in providing a good, and we therefore should be open to a variety of modes or combinations thereof.

Argument from participation in modes of provision

Many discussions of choice in economics focus on the sovereign consumers' choices. No doubt, market economies are very powerful engines of product choices and with it an improved standard of living. However, as a matter of lifestyle, increase in market choices are also viewed critically as costly, time-consuming, and to the detriment of sovereign individuals and communities (Paech 2012). A distinction is needed between a variety of products, and higher order choice for meeting ends. An example is the difference between a choice of cars, and the choice of meeting mobility needs (public transport, self-provision ... see Abraham 2015, p. 144).

Democratic states that foster and secure the dignity of their members should provide them with real choices regarding the mode of provision, not just simple order consumer product choice. This claim is supported by the argument from diversity above, but also by the central capability of participation, or what Nussbaum (2006) calls the political and material control over one's environment in choices that govern one's life. Real choice in such matters depends on the de facto availability of such modes of provision and the varieties of conversion they afford.

The diagnosis of current societies as consumer democracies indicates that such choices are limited, and in the extreme reduced to market provision. Monetary income or wealth is decisive, and hence there is a push to increase wage labour and monetary income so as to be able to afford consumption choices. Time for communal, public, or self-provision is short. Such factors in turn contribute to a cognitive frame in the focus of many post-growth critiques: that there is no choice but the market model as we know it; and inversely the call for emancipation (Gebauer 2018, p. 232) from such cognitive dependencies.

A choice in modes of provision is both instrumentally valuable (given the diversity of people and their contexts), and valuable intrinsically: it enables participation in structural choices that governs one's life and with it an improved freedom of entering and exiting social practices, or what Claassen calls navigational agency (Claassen 2017). Persons might not only choose to change modes (and associated market, communal and political practices), but also to "blend" them in new ways. For example, a civil society innovator might find cooperation with people from business beneficial for business planning and accounting. Thus the argument from provision choice supports a complex and collaborative pluralism of provision. It also is part of the explanation why democracy is essential for post-growth economies (Ott 2012).

Argument from the value of niches

A plurality of modes of provisions also offers a reservoir of ideas, skills and relations. Even if one way of doing things prevails in a context, it is helpful if the "mainstream" is tested by comparison with alternatives. For example, in a country like Germany much education is publicly provided, however, there is a "niche" for other providers. "Out of this niche for private, often non-compliant schooling reforming stimuli emerged, across the years influencing the public education system" (Preuß 2016, cited in Scheuerle et al. 2016, p. 199).

This niche argument is reinforced by a security consideration (Scheuerle et al. 2016, p. 199):

> Policies and regulations should allow for different models of financing and organisation. It is not only necessary to fund the creation of new ideas. It is also important to keep open the possibility to quickly implement good ideas in an emergency. This is an investment in the future, which allows for the societal stability and innovation capacity of society to come.

A plurality of modes of provision and with them of innovating is in a weak sense a fact across a variety of modern states. While one mode might be stronger than others in the respective regime, all modes of provision seem generally present. For example, the Soviet Union had a state-organised system of production; yet there were, already in the 1930s, publicly tolerated Kolkhoz markets for the exchange of artisanal production based on market prices (Fitzpatrick 1999, p. 57).[8] Likewise, even though market provision has significantly grown in capitalist regimes, household production has prevailed all along, making up a significant share of the economy (Offer 1997).

This observation suggests a presumption of equality as a global starting point for a stronger sense of appreciating the plurality of modes of provision: a priori, no mode is preferable as such to any other. Rather, each mode has its own advantages and disadvantages. The question is therefore the right "mix of modes" to improve agency possibilities and functioning outcomes, in particular with a view to the long-term stability and resilience of society. The post-growth economics critique of consumer democracy on this interpretation is a critique of an imperialist "commercialized" mode appropriating all other modes. The alternative is a Millian insistence on examining the practical truth of a mode by contrasting it with alternatives, thereby examining, refining and ultimately strengthening the conversion of goods to ends.

Blocked modes of provision

However, is such a presumption of equality not counter-intuitive, especially for post-growth economists and their frequent calls for de-commercialisation and de-privatization of post-growth-activists? Prima facie not in Paech's case, as there is a commercial area and with it also a place for commercial and private business work.

Still, this seems an insufficient reply, as the assertion of a commercialized area is not the same as supporting the provision of all goods in this area. In his defence of pluralism, Michael Walzer came up with an impressive list, when he argued that some elements of the social world should be blocked from exchange on the market: human beings, political power, criminal justice, basic liberties such as free speech, marriage and procreation rights, membership in the polis, exemptions from public service (military, juries etc.), political offices, desperate exchanges, prizes and honours, divine grace, love and friendship, and criminal sales (Walzer 1983, pp. 100–103).

Market exchange is a central bone of contention. But, generalizing Walzer, we also can ask whether a good should be provided informally, publicly, and/ or professionally. For example, in Greece, Germany, and the Netherlands home schooling is "blocked" by law, whereas it is legal in Hungary, France, and the UK. In many countries, the state provides education, but some argue it should only regulate it (see van der Linden, in press). The example indicates that there is controversy around "blocking."

This controversy refines the case for a plurality of modes of provision: where provision of a modes violates central capabilities and thus dignity, there is a case for blocking. This case emerges from a consideration of not just the market but all modes of provision. On the central capabilities conception advanced here, blocking should be considered in terms of the disadvantage(s) for central capabilities overcome (or created) via the respective provision.

Thus we have arrived at a refined sketch of the freedom case for a plural mode of provision. Moving from the consideration of individuals (in argument one) via specifically individuals in their roles as reflecting, deliberating, and decision-making participants (in argument two) to a Millian case for plurality to keep the practical truth of modes examined, a freedom case emerged and was refined with the inclusion of blocked modes. The case implies a shift from consumer society towards a plural economy. As Paech notes, subsistence is not a matter of "either/or but of more or less in a combined system of provision" (Paech 2012). This is a freedom case for a plural system of provision beyond a matter of subjective consumer preference only. It revises the classic liberal case for freedom as a matter of open markets and limited government in favour of a case for freedom across modes of provision.

Wage labour reduction and freedom

The case for plural modes of provision stated, we can re-examine the distinction at the heart of Paech's proposal between commercialized and de-commercialized areas and with it his call for a reduction of wage labour to a 20-hour week. Is this not an evident anti-freedom imposition? From a libertarian perspective, perhaps, an imposition that has travelled from social-democratic concerns of 20th century industrial states to the sustainability concerns of 21st century green government?

Historically, the reduction in work hours has resulted from social conflict and cooperation between the owners and managers of organizations, trade-unions, and the state. It suggests some evidence that there is a long-term development in the direction of distributing some productivity gains in favour of reduced wage labour.[9] In the light of such gains, Keynes (1930) already envisaged a 15-hour week 100 years from his time of writing, i.e. around 2030!

The freedom case cannot justify either the number 15, 20, or ... .[10] It is also not meant to downplay the political and cultural difficulties of a reduction of wage labour time in market economies where especially for the less well-off many questions of opportunity and status depend on wage labour (see Mock, this volume). However, it supports the case for a balanced starting point across modes of provision. 15/20 hours on this perspective are a regulative proposal for a structural default position, enabling freedom across modes of provision. A reduction in wage labour is justified as it promotes real opportunity to participate in communal and public provision.

Especially for public provision (as well as for regulation), it creates spaces for an omnipresent demand in late-modern democracies: participation. A demand that is often only simulated (Blühdorn 2013) as consumer citizens lack time and will to participate.

Moreover, the policy promises to reduce the threat of consumption reduction in a post-growth economy: it creates space, time, and legitimacy for fostering skills and relations for alternatives to commercial production and consumption. Thus the libertarian utopia of working hours resulting from contracts between individuals is replaced by a historically-embedded concern for fostering real choice given the societies we live in.

However, is such a policy even necessary? Will scientific-technological advance not free us both from economic toil and ultimately from any unsustainability challenge arising (as eco-modernists might argue)?[11] Already Keynes (1930) had noted the possibility of long-term "technical unemployment" as a result of technical change. The question leads to an important point. On one perspective, "the economic problem" is simply toil required to provide the means of subsistence – freedom is freedom from toil. In fact, Keynes's case for the 15-hour week is transitional: to "satisfy the old Adam" (or "protestant Max"), who needs to work as long as he has not learnt to simply enjoy himself. Once we have dealt with this culture lag, adapted habits and Protestant values, much less work than even 15 hours are needed and desirable. By contrast, on the perspective proposed here, the idea of labour as a technical, negative necessity is at best partial. There is freedom *in* the economic problem and the provision of subsistence. The example of care indicated the choices involved in receiving care, and caring for others, and with it relations and skills intricately woven into central capabilities of affiliation and control of the environment (to use Nussbaum's 2006 terminology).[12]

To be sure, we can imagine a techno-utopia where all care work is done by robots as a new kind of techno-professional provision following a robot code and rules of care. Old Adam can still care if he so cares – but he can also embark on new adventures. However, even in this case, old Adam requires structures to make such care options real options. Moreover, the utopia takes us to the limits of the current freedom argument and to the long-term consideration of limits of resources and emissions. A scenario of further techno-growth with "working slaves" – robots and other machines – requires energy, resources for production, etc. However, our starting point was precisely that we already transgress environmental limits. A recent survey shows that the case for green-growth and the absolute decoupling of economic growth is not met by evidence for resource inputs, and highly unlikely for a central resource output, i.e. greenhouse gas emissions, even in the most optimistic scenarios (Hickel and Kallis 2019).

In the old liberal case, the consideration of such boundaries was not important due to the promise of progress – the extension of freedom was associated with a belief in an ever-increasing standard of living for all (where this often included a disregard of other species). This promise, especially with its materialist connotation in the standard of living, is no longer plausible: sustainability of provisioning systems is called for (O'Neill et al. 2018). This is in part a case of "liberating the future from the affluence of the present," but really a more general issue of intergenerational and interspecies justice based on freedom but also on equality and solidarity: in recognition of the options of future generations; and in recognition of other living beings and their requirements of environmental space. Thus, the freedom case is important, but only a part of a larger case for a post-growth economy.[13]

Conclusion

The freedom case for a plurality of modes of provision strengthens domestic and communal provision, and thereby a movement in the direction of a more plural, post-growth economy. It makes a case for freedom as an important value, though by no means the only one, to respond to the Anthropocene, here referred to as the transgression of social and environmental planetary boundaries due to the growth of the global economy since industrialization (or, again, more playfully with Keynes (1930), since pirate Drake stole treasures on behalf of Queen Elizabeth I in 1580, allowing her to pay off debt and investing in the Levant company, thereby triggering a long-chain of accumulation and growth). It makes a philosophical contribution to the call for a renewed consideration of provisioning systems (O'Neill et al. 2018), and especially the social dimension of such systems.

While versions of "liberal rationalism" (see Jennings, Kish and Orr, this volume) might have played a liberating role in the past, they are too limited and therefore misleading as a guide for sustainability. The capability conception of freedom explored here affirms the importance of individuals. However, it also suggests to see individual agency on a continuum with patiency.[14] Liberty is important, but so are the vulnerabilities of community members as patients. Sovereign is she who knows her limits, and sovereign the society that is aware of collective and individual limits. This is a matter of accepting finitude in relation to consumption choices (Paech 2012, p. 129), and also of asserting choice options beyond consumption in markets only.

The last point also has an implication for potential strategies to facilitate more ecologically benign ways of living on the Earth (see Jennings, Kish, and Orr, this volume). The discussion above asserts a revised conception of liberty (vis à vis the old liberal view), and with it of freedom in an internally diverse economic space. Notable from this perspective is the emergence of a discourse on *social* innovation (Howaldt et al. 2018). The classic driver of growth in capitalism is no longer innovation tout court. Innovation has come to need an adjective, a qualifier (Ziegler 2017). Are social innovations the contemporary Gestalt of emancipatory innovation, breaking boundaries and searching new configurations of provision (or restoration of those marginalized in the present)? If so, society should develop a capacity to listen to them as messages, to hear them and to discuss them, especially where they challenge prevailing boundaries and push society into the direction of a more balance, plural economic space. This point takes us to further questions, beyond this chapter and perhaps better dealt with by political scientists, policy-makers, and activists: what institutions and democratic innovations are needed to foster and secure discussion and decision-making for plural modes of provision?

Notes

1 I would like to thank Gabriel Yaya Haage as well as the editors for their comments on an earlier draft. This chapter was written during a visit as guest professor at the Institut de l'EDDEC in Montreal. I would like to thank the IEDDEC for the hospitality and its commitment to advancing a different economy.
2 P. 130, own translation, based on the 2012, 9th edition.
3 See especially chapter VI for other visions.
4 The specification of a sufficientarian threshold is a subtle matter, not least as central capabilities are also relational in important aspects. For further discussion of this point see Ziegler 2020, chapter one.
5 For further discussion, see Ziegler 2020, chapter one.
6 Drydyk (forthcoming) captures this important aspect of dignity via the formulation of respect for the "vulnerable striving" of each person.
7 These arguments are taken from chapter four of Ziegler 2020.
8 Along with a "black" or "second economy" that always existed alongside state production.
9 Along with more recent, anecdotal evidence that reduction in labour time can also be a business case for employers (https://www.theguardian.com/world/2018/mar/29/heck-it-was-productive-new-zealand-employees-try-four-day-week, accessed September 27, 2018).
10 Paech sees the 20 as a rough estimate with an educational aim: stimulate debate (personal communication November 24, 2018)
11 On this discourse, see Cannavo, this volume.
12 See also Kish, this volume.
13 In a more recent book (Eppler and Paech 2016, 128), Paech in fact appeals to a global resource egalitarianism. Thus he would probably concede this point.
14 And raises further questions for research. For example, if sufficiency should be understood in terms of an ample minimum of central capabilities (as I have done here) or in terms of a social optimum (as Drydyk forthcoming argues).

References

Abraham, Y.-M. (2015). "La décroissance soutenable comme sortie du capitalism." *Possibles*, 4: 137–152.

Blühdorn, I. (2013). *Simulative Demokratie: Neue Politik nach der postdemokratischen Wende.* Berlin: Suhrkamp.

Claassen, R. (2009). "Institutional pluralism and the limits of the market." *Politics, Philosophy & Economics*, 8(4): 420–447. doi:10.1177/1470594X09345479.

Claassen, R. (2017). "An agency-based capability theory of justice." *European Journal of Philosophy*, 25(4): 1279–1304. doi:10.1111/ejop.12195.

Cohen, G. A. (1993). "Equality of What?: On welfare, goods, and capabilities." In M. C. Nussbaum & A. K. Sen (eds.), *Studies in development economics. The quality of life: A study prepared for the World Institute for Development Economics Research (WIDER) of the United Nations University* (pp. 9–29). Oxford: Clarendon Press.

Drydyk, J. (forthcoming). "Sufficiency reexamined." In F. Comim, S. Fennell, & P. Anand (eds.), *New Frontiers of the Capability Approach.* Cambridge: Cambridge University Press.

Eppler, E., & Paech, N. (2016). *Was Sie da vorhaben, wäre ja eine Revolution...: Ein Streitgespräch über Wachstum, Politik und eine Ethik des Genug.* München: oekom.

Evers, A., & Brandsen, T. (2016). "Social innovations as messages: Democratic experimentation in local welfare systems." In T. Brandsen, S. Cattacin, A. Evers, & A. Zimmer (eds.), *Nonprofit and civil society studies, an international multidisciplinary series. Social innovations in the urban context* (pp. 161–180). Heidelberg: Springer Open.

Fitzpatrick, S. (1999). *Everyday Stalinism: Ordinary Life in Extraordinary Times: Soviet Russia in the 1930s.* Oxford: Oxford University Press.

Gebauer, J. (2018). "Towards Growth-Independent and Post-Growth-Oriented Entrepreneurship in the SME Sector." *Management Revue*, 29(3), 230–256. doi:10.5771/0935-9915-2018-3-230.

Gorke, M. (2010). *Eigenwert der Natur.* Stuttgart: hirzel.

Hickel, J., & Kallis, G. (2019). "Is green growth possible?" *New Political Economy*, doi:10.1080/13563467.2019.1598964.

Howaldt, J., Kaletka, C., Schröder, A., & Zirngiebl, M. (eds.). (2018). *Atlas of Social Innovation. New Practices for a Better Future.* Dortmund: Sfs.

Kallis, G., Kostakis, V., Lange, S., Muraca, B., Paulson, S., & Schmelzer, M. (2018). "Research on Degrowth." *Annual Review of Environment and Resources.* Advance online publication. doi:10.1146/annurev-environ-102017-025941.

Keynes, J. M. (1930). *Economic Possibilities for our Grandchildren.* Reprinted in his *Essays in Persuasion* (pp. 358–373), New York: W. W. Norton & Co., 1963.

Nussbaum, M. (2006). *Frontiers of Justice: Disability, Nationality, Species Membership.* Cambridge: Harvard University Press.

O'Neill, D. W., Fanning, A. L., Lamb, W. F., & Steinberger, J. K. (2018). "A good life for all within planetary boundaries." *Nature Sustainability*, 1(2): 88–95. doi:10.1038/s41893-018-0021-4.

Offer, A. (1997). "Between the gift and the market: The economy of regard." *The Economic History Review*, 50: 450–476. doi:10.1111/1468-0289.00064.

Ott, K. (2012). "Variants of de-growth and deliberative democracy: A Habermasian proposal." *Futures*, 44(6): 571–581. doi:10.1016/j.futures.2012.03.018.

Paech, N. (2012). *Befreiung vom Überfluss: Auf dem Weg in die Postwachstumsökonomie.* München: oekom. English translation freely available online.

Scheuerle, T., Schimpf, G., Gänzel, G., & Mildenberger, G. (2016). *Report on Relevant Actors in Historic Examples and an Empirically Driven Typology on Types of Social Innovation.* Cressi Working Paper 29, SBS Oxford University Press.

Sen, A. (1999). *Development as Freedom.* Oxford: Oxford University Press.

Taylor, P. W. (1986). *Respect for Nature: A Theory of Environmental Ethics. Studies in Moral, Political, and Legal Philosophy.* Princeton, NJ: Princeton University Press.

Van der Linden, M. (2019 forthcoming). "The dutch school struggle: A historical case of creating economic space for freedom of education." In A. Nicholls & R. Ziegler (eds.), *Creating economic space for social innovation* (Chapter 8).

Walzer, M. (1983). *Spheres of Justice: A Defense of Pluralism and Equality.* New York: Basic Books.

Ziegler, R. (2017). "Social innovation as a collaborative concept." *Innovation: The European Journal of Social Science Research, 30*(4): 388–405. doi:10.1080/13511610.2017.1348935.

Ziegler, R., & Jacobi, N. V. (2018). *Fair (Economic) Space for Social Innovation? A Capabilities Perspective* (Cressi Working Paper Series No. 44). Oxford, pp. 1–28.

Ziegler, R. (2020 forthcoming). *Innovation, Ethics and Our Common Futures. A Collaborative Philosophy.* Cheltenham, UK: Edward Elgar.

11

COGNITIVELY UNSTABLE RATIONAL AGENTS

A new challenge for economics in the Anthropocene?

Morgan Tait

Introduction

Despite concerted efforts on the part of its critics, neoclassical economics[1] retains a special authority in both practical governance and the academy. This authority is peculiar to neoclassical economics, excluding even other economic approaches such as ecological economics. It reflects the curious intellectual history of the neoclassical tradition. The curiosity lies in the historical source of neoclassical authority in the normative assumptions of rational choice theory, and its application to the ontology of what game-theorists call "non-parametric game environments," where acting rationally depends on anticipating and reacting to the often complex behaviours of other rational agents as they anticipate and react to you.[2] Yet given the pretensions of neoclassical economics to scientific as well as social legitimacy, it is ironic that the judicious application of rational choice theory calls into question the rationality of the ontological assumptions of neoclassical economics, and in particular the assumption that the enclosure of the commons by the market is a kind of positive-sum game in a non-parametric social system whose boundaries are essentially unknowable or "unsurveyable" (Hayek, 1988).[3]

Economists have often been accused of physics envy, and not without cause. Physicists use an inferential method with a proven track record of epistemic utility; believing what physicists say about the natural world is obviously rational in exactly the sense that economists would have us be rational. Physicists also contributed to the modern formalization of the discipline of neoclassical economics during the great depression, when there was a felt need for their intellectual talents in the adjudication of public affairs. The new economic formalism was particularly indebted to equilibrium thermodynamics and classical mechanics (Mirowski, 1992, Koopmans, 1979), while novel foundational contributions were also made by mathematical physicists collaborating with economists (e.g., von Neumann, 1945–1946; von Neumann and Morgenstern, 1944). The Anthropocene forces a second look at these intellectual roots of economics. It is an epoch in which the physical sciences must once again intrude into the adjudication of public choices, particularly in domains where the choices that agents make have catastrophic physical repercussions. A discussion of the ontology of neoclassical economic systems can and must be had, on terms that both the neoclassical economist and the climate scientist are prepared to grant. The natural scientists once again have cause to turn their attention to the activities of economists, and, just as they did in the 1930s, they are in a position to radically reorient the discipline from within, although for very different reasons.

Why hasn't this transformation already taken place? Why do neoclassical economists continue to have disproportionate clout in social affairs? This phenomenon can be traced to two foundational assumptions that resonate deeply with the core values of the liberal humanist tradition, about which the natural sciences have until recently remained relatively silent.

The first is a commitment to a kind of *platonic ideal of rational agency*, derived from decision theory. While championing rationality as an instrumental virtue, the economist can dodge charges of descriptive inaccuracy by retreating to a normative account of how human beings "ought" to act, for example by maximizing their utility, regardless of how they *actually* act. This value commitment is analogous to the myth of the social contract, in that its force is normative rather than descriptive. As we will see, this commitment to instrumental rationality is also a core value commitment of the natural sciences, despite its non-empirical flavour. This has been made explicit by philosophers of science but is largely tacit among practicing scientists.

The second foundational commitment is to a principled *metaethical value-agnosticism*. As Pareto acknowledged, "One is grossly mistaken ... when he accuses a person who studies *homo economicus* ... of neglecting, or even scorning, *homo ethicus*" (Pareto, 1906). The economists do not tell us what to value or coerce us into changing our values, it is argued; instead, given our values, the economist supplies an account of how to go about exercising them, whether by sharpening our own decision-making, or incentivizing others to change their patterns of behaviour. This second commitment plays on the democratic norms of tolerance and freedom from coercion and is largely aspirational inasmuch as actual economists are not, and do not pretend to be, value-neutral.

These two ideological commitments are highly resilient, persisting despite compelling arguments that *homo economicus* is a fiction, and perhaps unlike the social contract, a morally bankrupt and ecologically unsound one. These commitments are sufficiently strong to withstand criticisms from other social sciences and the humanities, which do not command the same respect as economics in public disputes. Yet the discourse of the Anthropocene offers a new kind of challenge. It implies a new ontology that exposes a tension at the core of neoclassical economics and renders that tension untenable. Rather than charging that economists are bad anthropologists, or bad ethicists, the new discourse attacks the discipline in relation to the method of the natural sciences, which employ epistemic methods dear to all economists. Driven by a sense of planetary emergency, these disciplines have only recently turned their attention to social systems. These systems are no doubt "non-parametric" in the game-theoretic sense, and therefore arguably must be studied on those terms (Ross, 2019).[4] Yet the Anthropocene invokes an ontology which places parametric boundaries around all non-parametric environments. This chapter argues that any economist who believes in the neoclassical ideal of "enlightened self-interest" should take heed on the grounds that it would be irrational not to. Hence, this chapter attempts to cast the Anthropocene, and the corresponding "anthroposphere" (Baskin, 2015) as a radical ontology, one capable of shedding light on "what is ontologically different about ecological economics" (Spash, 2012). By showing how neoclassical economics depends on the natural sciences for its credibility as well as to define the absolute boundaries of economic discourse, the present chapter will make the case that in the anthroposphere, *homo economicus* is a casualty of its own ideological commitments.

Physics envy in the Anthropocene

In "Do economists suffer from physics envy?" Philip Morowski writes that "when you've got physics envy and got it bad, no model will ever gain substantial allegiance in neoclassical economics until it has first earned its spurs in physics" (Mirowski, 1992, p68). The Anthropocene has indeed earned its physics spurs, though it has yet to capture the attention of most economists. Yet the methods used by climate scientists are consistent and continuous with those of the more theoretical and foundational sciences of physics and chemistry. This continuity is rooted in the epistemic commitments shared by all natural sciences and, ironically, by neoclassical economics.

What makes the sciences rational from an economic perspective?

The real root of physics envy among economists is the paradigmatic rationality and explanatory reach of the idealized physicist. Above all else, the economist needs to be trusted in the domain of public affairs, where the actions, interests, and incentives of agents matter, and she rightly regards the rational method of the natural sciences as the ideal of epistemic trustworthiness: we trust engineers trained in physics to build bridges that will not collapse. In the public imagination, this trustworthiness is related to spectacular successes in predicting and explaining natural phenomena, engendering claims that are testable and falsifiable, unifying our experience and suggesting new avenues of empirical investigation, and generating reproducible and reliable interventions into the natural world. But what explains the capacity of the natural sciences to achieve these ends? Here the economist is right to focus on what makes the ideal physicist paradigmatically rational: the epistemic utility of thinking like a physicist.

To begin with, the neoclassical economist will note that the concept of expected utility is useful for characterizing how we ought to act in contexts of uncertainty of any kind. For example, if I am deciding whether to bring an umbrella when I leave the house, I will have to consider the utility of staying dry, the disutility of getting wet or carrying an umbrella, and my assessment of the likelihood that it will rain. The expected utility of bringing an umbrella will then determine, based on my preferences, whether it is rational to bring an umbrella when I leave the house. Epistemic utility generalizes this idea in order to characterize whether it is rational to hold certain beliefs. For example, I might wonder whether my assessment of the probability that it will rain is reasonable. How can I go about answering this question?

We can do so by appeal to the "epistemic state" of the agent, which can be thought of as a characterization of her belief commitments, and their relation to states of the world (Pettigrew, 2015). We can then ask, what is the value of being in a given state of belief, given the state of the world? Should I believe that the probability of rain is 50%? If we think in purely instrumental terms, or in terms of means and ends, the most epistemically valuable belief states will be those that allow us to most efficiently achieve our goals, whether in relation to other agents or to a non-agentic physical environment. In the latter case, believing that it will rain with a 50% probability is more valuable than believing it with any other probability, assuming that I attach some non-zero utility to rain, and that the chances of rain are indeed 50%.

How does one determine what credence, or degree of belief, to attach to any given proposition, such as that it will rain? The standard method in measurement theory, which is foundational in discussions of confirmation theory in the philosophy of science, is to appeal to expectation values, which are closely related to expected utility. If I want to know whether a die is fair, I should toss it as many times as I can, bearing in mind that the average outcome should converge to 3.5, the expectation value of a fair die. Repeatable experiments of this kind form the evidentiary basis of the natural sciences and provide a natural basis for interpreting which beliefs about natural phenomena are most rational to have, namely those which match up, in the long run, with the revealed outcomes of experimental interventions into the world. Scientific hypotheses are just our best attempts to explain the results of these repeated experiments, and if we are lucky, to predict the results of experiments that have yet to be performed. The rational agent, on this view, will update their beliefs about the natural world by conditionalizing, using Bayes' rule, on the evidence provided by experiment. In this way, the epistemic state converges on an accurate representation of the state of the world through experimentation, or equivalently, through accumulation of evidence, along with the updating of belief in accordance with the demands of the probability calculus.

The incredible fact about our natural scientific interventions into the world is that they have made it rational to believe in an ontology of mathematical natural laws. The staggering epistemic utility of belief in the existence of natural laws is the real source of physics envy. From the point of view of epistemic utility, the mathematical physicist is wealthy indeed (though she may have fewer dollars than her economist colleagues). The imputation of parsimonious natural laws to the world allows rational agents to systematically generalize by abstracting away from those details of experience that can be regarded as irrelevant from a nomological point of view. There is great cognitive value in being able to filter incoming information for salience; for example, the details of bridge-building from a physical point of view do not change whether the bridge is to be built in Copenhagen or San Francisco, or even on the moon. By allowing us to systematically ignore what does not matter, and identify real and universal patterns in natural phenomena, our belief in mathematical laws allows us to intervene in the world with astounding efficacy and efficiency. Of course, there is no *a priori* reason to expect the world to be amenable to such a mathematical treatment; the fact that natural laws seem to exist at all is our good fortune. It is a conceit of physics, rooted in remarkable experimental successes and explanatory unification, that we will continue to find natural laws if we develop the mathematical techniques to look for them. A similar conceit has had disastrous consequences in empirical economics where there are no such laws.[5]

Anthropocene and anthroposphere: Modelling agency, agentic models

The paradigmatic rationality of the ideal physicist is the subject of envy among economists, but also constitutes an existential threat to the discipline when that rationality is applied to the empirical observation of the ecological impacts of human behaviour. In this way the Anthropocene draws attention to a core epistemological tension in enlightenment liberalism, inherited from its Greek and Medieval ancestry, between humanistic and naturalistic currents in philosophical and scientific discourse. Humanists, whose intellectual ancestry traces back at least to Plato, want to shape the world according to ideals that transcend the actual world in order to provide a basis for judging it; naturalists, following Aristotle, rightly point out that correctly describing the world itself has great value whatever our ends. One way to negotiate this tension is to view the Anthropocene through the lens of pragmatist philosophy of science, taking as inspiration an idea expressed by Hilary Putnam and others: that dualities in our experience do not imply the experience of duals (Putnam, 1982). In the case of the anthroposphere, the ontology corresponding to the Anthropocene epoch, the dualities in question are the humanistic and the natural scientific, and the temptation Putnam's aphorism would have us avoid is a dualism of natural and human ontologies. Yet precisely because it sometimes seems to lean on such a dualism, or at least retreat to it when it is under critical assault, neoclassical economic theory is especially threatened by the Anthropocene. Economists do not want their discipline to be naturalized, even though they acknowledge the epistemic authority of the natural sciences and may count themselves as naturalists. In this way the neoclassical economist is both a humanist and a naturalist, a dual persona that is unstable in the integrated ontology of the Anthropocene.

It is tempting to avoid dualism via reduction. Natural scientists might reason that like evolutionary theory, heliocentric astronomy, or quantum mechanics, the reality of the Anthropocene compels us to accept as real an uncomfortable but entirely naturalistic epistemology and corresponding ontology of limits to growth, regardless of the social and political consequences of doing so. This is the approach adopted by ecological economics. Critical theorists balk at this idea, arguing that the Anthropocene is a political construct intended to justify technocratic managerial interventions. This critical approach effectively reduces the duality to a social ontology. The neoclassical economist, like most lay people, is not comfortable with either reduction: like all humanists, she wants to believe in the power of human reason to transform the physical world according to rational needs. But she also wants to argue that the method of the sciences is the paragon example of the exercise of human reason, even if the world it describes makes reason a product of natural processes rather than an agent of their arbitrary transformation. This tension remains tenable only as long as the natural world described by the sciences can accommodate the conceits of humanism; if it cannot, either the humanism or the natural ontology which threatens it will have to give.

Adopting the Anthropocene as both a natural and political concept requires that we understand the causal processes involved in bringing it about as physical-political-economic processes, while also viewing those processes as objectual, parametric, identified by replicable research methods, and generative of public consensus. The latter task, of marshalling public goodwill, suggests a circumscribed role for economics. The social causes traditionally investigated by economists and other social scientists are not, however, easily subjected to such a naturalistic treatment. The main thesis of this paper is that the marriage of the anthroposphere with the rational choice theory presupposed by neoclassical economics is in fact cognitively unstable. Rational choice theory compels us to adopt an ontology that collapses non-parametric environments, the special objects of study for neoclassical economics, into objects of *physical* study. In light of this instability, economists face a dilemma: embrace the Anthropocene, and give up on *homo economicus* construed as an ideal rational agent in an "unsurveyable" environment, or retain the tenets of traditional economics by denying the evidence for an Anthropocene. Yet if the economist wants to go the latter route, they will have to explain how something as rational as the scientific method can give rise to an ontology that is to be rejected.

The two faces of homo economicus: Neoclassical economics as cognitively unstable?

The previous discussion has suggested that *homo economicus* is a hybrid being, pulled in incompatible directions by its own nature. In the first place, *homo economicus* is a kind of normative platonic ideal of rationality, construed in the economic sense of self-interest and utility maximization, strategically negotiating non-parametric environments, characterized in large measure by the actions and interests of other agents. *Homo economicus* is a rational reconstruction of human agency rather than a historical or anthropological account of human nature. On the other hand, *homo economicus* is also ostensibly the product of *empirical* inquiry into the question of how in fact to navigate those environments where the attitudes and motivations of other agents matter to you, and yours to them. For example, *homo economicus* is the sort of being that has incentives, and collections of such beings can be "incentivized" with the right sorts of economic levers. *Homo economicus* embodies the kind of rationality that is paradigmatic of the ideal experimental physicist, when the "physicist" is forced to play games.

Two traditional threats to neoclassical economic hegemony, and two strategies for evading them

Forceful critiques of neoclassical economics have been on the offing for as long as economics has been the dominant social science in policy and governance circles. Yet despite their apparent cogency, these arguments have made remarkably little headway in reducing the hegemony of the discipline. I begin by outlining these argumentative strategies before moving on to assess their efficacy in relation to the Anthropocene. Two general strategies can be identified.

One common approach is to point out that economics systematically fails to accurately describe or identify real features of human nature, both at the individual and the social scale, and therefore does not qualify as a genuine descriptive science. This sort of argument has been levied by social theorists of all stripes. To take just two examples, ecological critiques of traditional economics stress the falsity of the assumptions of unlimited growth and substitutability of natural and social capital (Daly, 1993). At the scale of individual agents, Nobel prize-winning psychologist Daniel Kahneman has argued that the notion of rationality implicit in expected utility theory, one of the conceptual cornerstones of neoclassical economics, does not correspond with psychological reality. If we distinguish between "experienced utility," the amount of actual hedonic pleasure we experience as a result of our choices, and "decision utility," or the "wantability" of our choices, it turns out that the two notions often don't coincide in real agents. The "cold hand" study shows that human subjects routinely prefer choices that do not maximize their utility in the hedonic sense (Kahneman, 2011). As Kahneman puts it, "the evidence presents a profound challenge to the idea that humans have consistent preferences and know how to maximize them, a cornerstone of the rational-agent model" (Kahneman, 2011, p385).

Another consistent argumentative strategy, especially among environmentalists and ecologists, is to criticize economics on ethical grounds. Here there are many cogent points to be made. It seems unreasonable on ethical grounds to try to assimilate or reduce all values to preferences in order to make utility calculations. The worry may be that marketized systems of valuation are exploitative, or corruptive of non-market goods, or both (Kelman, 1981, Sandel, 2013, Masood, 2018). Relatedly, it is arguably necessary on ethical grounds to distinguish instrumental from intrinsic values, and to avoid trying to reduce the former to the latter (Malm and Hornborg, 2014). Further, it is often argued that the economic framework of capitalism or neoliberalism is inherently unsustainable, leading to the systematic exploitation of the environment and of marginalized communities, the proliferation of exploitative power structures and perverse incentives, and the offloading of disproportionate costs onto future generations and the less powerful among the living (cf. Jamieson, 1992, Foster et al., 2010, Hamilton et al., 2015).

Both the descriptive and ethical arguments mentioned above have real cogency, and their failure to undermine the credibility of mainstream economics is telling. It is not simply a matter of vested interest or power propping up received economistic dogma, although unequal distributions of power and the privilege that they entail no doubt contribute to political and intellectual inertia. As mentioned in the introduction, there are two normative ideas germane to economics that have great rhetorical power. The first is that economics is rooted in and legitimated by the normative epistemology of decision theory. To be rational is the highest regulative ideal, since irrational behaviour can and will be systematically exploited either for nefarious purposes by conscious exploiters such as Dutch bookies, or incidentally by natural selection. Furthermore, the achievement of our goals is made possible by our rational assessment of available information, its consequences for us, and our assessment of the instrumental value of various goods in achieving our ends whatever they may be. If someone should complain that actual humans are not rational utility maximizers, the retort is that actual humans are not ideal moral agents either and that the articulation of a normative ideal of conduct is not undermined by the banal observation that few actual agents live up to such a standard. It isn't that humans *are* rational, it's that they *ought* to be, and economics, building on the insights of expected utility theory, can explain what this means in practice, for example, in the market.

Following on this idea is the related notion that reason is a supreme instrumental good, but *only* an instrumental good. The ends to which rational agency is directed, the argument goes, are not for the economists to decide. This supposed value agnosticism coheres very nicely with the liberal doctrine of tolerance.[6] It also makes it possible to value goods in a political setting without presuming that everyone will share the same values. Take, for example, the valuation of "ecosystem services." Perhaps you believe that certain goods provided by ecosystems are in fact intrinsic goods, worthy of protection regardless of their instrumental benefit to humans. That's fine, says the economist: go ahead and believe in such things. They may even exist. Yet so long as we can monetize intrinsic goods, we can get around the thorny political problem that we may not agree about what has intrinsic value by letting the market mechanism determine extrinsic value at the margin, in aggregate. And if you think that something crucially important is left out of such a metric, you can communicate your conviction to those who do not share it through your willingness to pay for holding it. Something like this attitude is evident in Costanza et al. (1997), which presents the following argument:

> A frequent argument is that we should protect ecosystems for purely moral or aesthetic reasons, and we do not need valuations of ecosystem services for this purpose. But there are equally compelling moral arguments that may be in direct conflict with the moral argument to protect ecosystems; for example, the moral argument that no one should go hungry. Moral arguments translate the valuation and decision problem into a different set of dimensions and a different language of discourse; one that, in our view, makes the problem of valuation and choice more difficult and less explicit. But moral and economic arguments are certainly not mutually exclusive. Both discussions can and should go on in parallel.
>
> *(Costanza et al., 1997, p255)*

In other words, moral arguments are fine, but the problem of reconciling different moral opinions, or moral trade-offs, is politically intractable. Furthermore, we can and do monetize things like human life when we cost things like bridge maintenance and car safety. Costanza et al. are not arguing that the value of a human life is exhausted by such costs, but it is a mistake to argue on this basis that we cannot or should not place a monetary value on human life. And the same goes for the other things we value by extension.

The rhetorical strategy of the economist when confronted with resistance on moral grounds is thus to plead political expedience and moral neutrality and stress the pragmatic value of econometric approaches to valuation. On the other hand, when attacked on descriptive grounds, the economist will invoke the platonic ideal of rationality, arguing that given the existence of utilities, there is a determinate answer to the question, "how ought we to behave so as to maximize wellbeing?" Both rhetorical strategies are at play in documents such as the Millennium Ecosystem Assessment, which speaks of "market failures and the misalignment of incentives" as a challenge to be overcome in reversing the degradation of ecosystems, exemplifying the relatively recent proliferation of the language of incentives in steering or managing large-scale social change projects (MEA, 2005; Sandel, 2013).

Conclusion: Aristotle's revenge: A new ontological threat to neoclassical economics?

The discourse of the Anthropocene presents a profound new challenge to this economic hegemony, one which is arguably implicit in many existing critiques of the discipline, but which is made explicit by the integration of natural and social systems implied by the marriage of nature and humanity in the word

"Anthropocene." The truly radical feature of the Anthropocene epoch, from a philosophical point of view, is that it immediately presents itself as a hybrid epoch whose ontology is both natural and social, yet reducible to neither category. The Anthropocene is hotly contested, despite its widespread adoption as a term of use, because it is very difficult to approach environmental governance without collapsing into either the natural scientific or the social scientific explanatory mode (Tait, 2019). Yet neither mode alone is up to the task of understanding the ontology it describes.

In this troubled epistemic-explanatory landscape, economics faces a unique challenge as a discipline that purports to be both scientific and humanistic. The ontology described by the economist is non-parametric, yet the anthroposphere collapses the game-theoretic analysis of markets into the realpolitik of ecological emergency.[7] It is as if the economist is describing poker strategy to a rapt audience in the parlour of the Titanic. Adam Smith's notion that markets are characterized by competition between rational agents engaged in positive-sum interactions (and hence growth), in an essentially non-parametric and indeed unknowable environment, is plausible so long as the parametric "externalities" implied by such market systems are themselves unknowable.[8] Given our current state of knowledge, more centralized, less distributed governance frameworks than that implied by (free) markets are rationally required.

What is radical about the new ontology of the Anthropocene is precisely that it compels economics to treat humanity in causal descriptive (parametric) terms, and to get the actual description exactly right. The fatal irony for the neoclassical economist is that the sort of rational belief commitments that support the anthropocenic worldview are the very same rational commitments that she holds up as the ideal for all rational agents. In the Anthropocene, we cannot take seriously the idea that neoclassical economics describes a platonic ideal of rationality for evaluating human behaviour, while also denying the existence of limits to economic growth, and as recognized by the Aristotelian philosophical tradition, to our capacity to idealize away from the physical circumstances in which human beings find themselves. Limits to growth, but also limits to agency, are implied by the platonic ideal of rationality, inasmuch as such rationality is exercised by the natural scientists purporting to show that we live in an era of unprecedented human influence on the biosphere. It is characteristic of the Anthropocene that human ideas, including the ideas of neoclassical economists, have real ecological impacts which we ignore at our own peril. Hence the neoclassical economist faces a stark choice given the cognitive instability of their belief commitments: either concede that intentionality and agency must be described in naturalistic terms, and described accurately, thereby cutting off the standard normative retreat to humanistic scruples and game-theoretic ontologies outlined above; or reject their own intellectual roots and deny the ontology of the "anthropocenologists" (this term is due to Bonneuil and Fressoz, 2017).

Perhaps most neoclassical economists would choose the second option, following Julian Simon's lead (Simon, 1981). One needn't believe in libertarian free will to believe that human agency is irreducible to the language of the natural sciences; one need only be a game theorist. It is one thing to deny the existence of libertarian free will, as many economists no doubt would, and quite another to insist on describing human beings in the naturalist/objectivist mode demanded by the anthropocenologists. To do the latter, according to Simon et al., is to ignore the fact that the economy is a non-parametric game environment, where reasons for action are crucial to understanding human agency and hence also to understanding the overall dynamics of the system.[9] Along similar lines, the philosopher Thomas Nagel writes that

> the application of certain concepts [of reason] from inside overpowers the attempts to grasp that application from outside and to describe it as a finite and local practice. It may look small and "natural" from the outside, but once one gets inside it, it opens out to burst the boundaries of that external naturalistic view
>
> *(Nagel 1997, pp71–72)*

Nagel could well have been describing the rhetorical power of rational choice theory, even in the face of ecological or other physical constraints, and even when those constraints receive their assertability from rational choice theory itself. As I have argued above, this is one of the chief seductions of neoclassical economics, that it treats instrumental reason as a supreme good, and also a crucial lens through which environments containing rational agents must be understood in order to achieve our own ends.

If it is no longer rational to believe the deliverances of rational choice theory, given our preferences, then we ought not to. Speaking on the issue of treating human beings as natural objects, and whether one needs libertarian free will to adopt "reactive attitudes" of praise and blame in moral judgment, P.F. Strawson writes:

> Quite apart from the issue of determinism, might it not be said that we should be nearer to being purely rational creatures in proportion as our relation to others was in fact dominated by the objective attitude [treating them as natural objects]? I think this might be said; only it would have to be added, that if such a choice were possible, it would not necessarily be rational to choose to be more purely rational than we are.
>
> *(Strawson 1962, p20)*

Strawson's point is well taken. Relating to other agents in a manner dominated by the "objective attitude," by treating them as natural objects, may indeed be rational, but it is an empirical question whether this is so. As Strawson stresses, it does not immediately follow from its being rational, that treating others in such a fashion would be to our collective benefit. Perhaps treating others as moral agents, or as (fictional) rational utility maximizers, serves an important social function. For example, a system of retributive justice might ultimately be justified on utilitarian grounds even if individual actors are constrained to act according to natural causes and don't really "deserve" to be punished for their transgressions in any deep sense. So it does not follow from the rationality of Strawson's "objective attitude" alone that we should choose to be rational, precisely because choosing to do so might be to our collective detriment. But if our failure to adopt an objective attitude is responsible for the destruction of the planet, for ourselves as well as for other species, then treating human beings as natural objects is, paradoxically from the point of view of the neoclassical economist, the *most* rational strategy to play.

Notes

1 By "neoclassical economics" I have in mind the cluster of ideas associated with the dominant 20th century "neo-classical-Keynesian synthesis," which presumes that agents are rational utility maximizers, motivated by (enlightened) self-interest to express rational preferences. According to Daly's (1993) definition, these ideas are supplemented by the claim that the economy is to be treated as an isolated system in the thermodynamic sense. The term "neoclassical" should be contrasted with "classical" economic theories associated with writers like Adam Smith and Jeremy Bentham.

2 Thus Ken Binmore (1994) writes: "A person would be stupid in seeking to achieve a certain end if he ignored the fact that what other people are doing is relevant to the means of achieving that end. Intelligent people will coordinate their efforts to achieve their individual goals."

3 Thus Hayek writes: "[Adam Smith] was the first to perceive that we have stumbled upon methods of ordering human economic cooperation that exceed the limits of our knowledge and perception. His 'invisible hand' had perhaps better have been described as an invisible or unsurveyable pattern."

4 Quoting Ross 2019: "[T]he aspect of the environment that is most important to the agents' achievement of their preferred outcomes is the set of expectations and possible reactions to their strategies by other agents. The distinction between acting *parametrically* on a passive world and acting *non-parametrically* on a world that tries to act in anticipation of these actions is fundamental." This distinction is also fundamental to understanding the legitimacy and authority of economics, which attempts to empirically model, predict, explain, and influence such game-theoretic environments.

5 For example, the claim that as markets become more competitive, they become more efficient, is false when expressed as a general law.

6 As does the doctrine of the unknowability of markets, which suggests an argument against centralized economic planning that coheres well with liberal values of freedom from coercion, etc.

7 For example, Paul Crutzen (2002) writes: "A daunting task lies ahead for scientists and engineers to guide society towards environmentally sustainable management during the era of the Anthropocene. This will require appropriate human behaviour at all scales, and may well involve internationally accepted large-scale geo-engineering projects, for instance to 'optimize' climate."

8 Quoting Smith: "The statesman who should attempt to direct private people in what manner they ought to employ their capitals, would … assume an authority which could safely be trusted, not only to no single person, but to no council or senate whatever" (Smith 1994, p 485). Indeed, the unknowability of the market suggests that free action on the part of economic agents is paramount to the generation of wealth, a fact

that coheres all too well with libertarian political sentiments. Yet markets embedded in larger ecologies are subject to knowable global constraints in ways that Smith did not anticipate, that require coordinated governance efforts that many economists unhelpfully dismiss as "inefficient."

9 I submit that this is the most charitable way to understand Simon's claim that natural resources are "infinite."

References

Baskin, J. (2015). "Paradigm dressed as epoch: The ideology of the anthropocene." *Environmental Values 24*: 9–29.

Binnmore, K. (1994). *Game Theory and the Social Contract Volume I: Playing Fair*. Cambridge, MA: MIT Press.

Bonneuil, C., & Fressoz, J. (2017). *The Shock of the Anthropocene*. New York: Verso.

Costanza, R., d'Arge, R., de Groot, R., Farber, S., Grasso, M., Hannon, B., ... van den Belt, M. (1997, May). "The value of the world's ecosystem services and natural capital." *Nature, 387*: 253–260.

Crutzen, P. (2002). "Geology of mankind." *Nature, 415*: 23.

Daly, H. (1993). "Steady-state economics: A new paradigm." *New Literary History, 24*(4): 811–816.

Foster, J. B., Clark, B., & York, R. (2010). *The Ecological Rift: Capitalism's War on the Earth*. New York: Monthly Review Press.

Hamilton, C., Gemenne, F., & Bonneuil (eds.) (2015). *The Anthropocene and the Global Environmental Crisis: Rethinking Modernity in a New Epoch*. London: Routledge.

Hayek, F. (1988). *The Collected Works of Friedrich August Hayek*, ed. William Warren Bartley. Chicago: University of Chicago Press.

Jamieson, D. (1992). "Ethics, public policy, and global warming." *Science, Technology and Human Values, 17*(2): 139–153.

Kahneman, D. (2011). *Thinking, Fast and Slow*. Toronto: Doubleday Canada.

Kelman, S. (1981). "Cost-benefit analysis: An ethical critique." *Regulation, 5*: 74–82.

Koopmans, T. (1979). "Experiences in moving from physics to economics." Unpublished talk delivered to the American Physical Association, New York 29th January 1979. Copy deposited in Koopmans Papers, Sterling Library Archive, Yale University, Box 18, Folder 333.

Malm, A., & Hornborg, A. (2014). "The geology of mankind? A critique of the Anthropocene narrative." *The Anthropocene Review, 1*(1): 62–69.

Masood, E. (2018). "Battle over biodiversity." *Nature, 560*: 423.

Millennium Ecosystem Assessment: Ecosystems and Human Well-Being. (2005). Retrieved from https://www.millenniumassessment.org/documents/document.356.aspx.pdf

Mirowski, P. (1992). "Do economists suffer from physics envy?" *Finnish Economic Papers, 5*(1): 61–68.

Nagel, T. (1997). *The Last Word*. New York: Oxford University Press.

Pareto, V. (1906). *Manuale di Economica Politica*. Societa Editrice Libraria, Milan.

Pettirgrew, R. (2015). "Epistemic utility arguments for probabilism." *The Stanford Encyclopedia of Philosophy* (Spring 2016 Edition), Edward N. Zalta (ed.), Retrieved from https://plato.stanford.edu/archives/spr2016/entries/epistemic-utility/.

Putnam, H. (1982). "Three kinds of scientific realism." *Philosophical Quarterly, 32*(128): 195–200.

Ross, D. (2019). "Game theory." *The Stanford Encyclopedia of Philosophy* (Spring 2019 Edition), Edward N. Zalta (ed.), Retrieved from https://plato.stanford.edu/archives/spr2019/entries/game-theory/.

Sandel, M. (2013). *What Money Can't Buy*. New York: Farrar, Strauss and Giroux.

Simon, J. (1981). *The Ultimate Resource*. Princeton, NJ: Princeton University Press.

Smith, A. (1994). *The Wealth of Nations*. Ed. Edwin Cannan. New York: Modern Library.

Spash, C. L. (2012, May). "New foundations for ecological economics." *Ecological Economics*, 77, 36–47.

Strawson, P. F. (1962). "Freedom and resentment." *Proceedings of the British Academy*, 48: 1–25. Reprinted in Fischer and Ravizza, 1993.

Tait, M. (2019). "Should naturalists believe in the anthropocene?" *Environmental Values*, 28(3): 367–383.

Von Neumann, J. (1945 [1946]). "A model of general equilibrium." *Review of Economic Studies*, 13: 1–9.

Von Neumann, J., & Morgenstern, O. (1944). *Theory of Games and Economic Behavior*. Princeton: Princeton University Press.

12

THE CIVILICENE AND ITS ALTERNATIVES

Anthropology and its longue durée

Joshua Sterlin

We do not live in the Anthropocene, but rather, the period of time defined by agricultural civilization. If we are to take this definitional question seriously, then the implications are immense for our understanding of not only our ecological predicament, but the entire narrative of our species, and therefore its hopes for future survival. Accepting the fact that our present era of mass extinction and climate change are the result not of something necessarily inherent to *Homo sapiens* as a species, but rather of a particular style of living and notion of a particularly singular human liberty, constrained to certain historical cultures, opens up space to consider already existent alternatives to what I am terming the *Civilicene*. It is indeed the fact that any other way of life, and conception of self, is framed as an "alternative" (Gibson-Graham 2008) which further reifies and performs the dominant position of the Civilicene narrative. If we are to de-centre our current way of life as teleologically inevitable, we must overcome this framing, allowing the expansion of both our cultural conception of the human being, and possible ways of life and nature–culture relations.

Anthropology's *raison d'être* has in some sense always been the study of these "alternatives" and their Western construed "vulnerability" (which is itself a form of othering). The reason to turn to this discipline specifically in this time has been aptly summarized by Graeber (2004), who wrote that we must "look at those who are creating viable alternatives, try to figure out what might be the larger implications of what they are (already) doing, and then offer those ideas back, not as prescriptions, but as contributions, possibilities – as gifts" (p.12). This has not always been the orientation of the discipline, being itself steeped in Western notions of cultural superiority. However, the drive to compare the vast diversity of humankind has always been at the centre of anthropology. If indeed each culture, as Wade Davis described it, is a unique answer to the question of "what does it mean to be human and alive?" (2009) then with each loss, each extinction of language and lifeways, we are as a species not only grieved and mourning, poorer of spirit, but as well *less resilient* in a way similar to an ecosystem experiencing untold extinctions. Davis further reminded us of the intimate connection of these two losses, of ecocide and ethnocide, and how the loss of cultures is to the mind what the loss of old growth forests is to the ecology. What the ethnographic record shows us are possible worlds, and notions of both human and non-human agency and its attendant liberties, both in that they are existent, and that they are *attainable*.

Anthropological survival

The ascendance of this as-of-now not yet geologically accepted, though certainly widely culturally recognized, marker of our commonly created geologic era has not turned any real public attention to the thoughts of those who study its protagonist, the *anthropos*, in full. There is good reason for this as Graeber (2004) aptly pointed out in *Fragments of an Anarchist Anthropology*: "anthropology seems a discipline terrified of its own potential. It is, for example, the only discipline that actually takes all of humanity into account, and is familiar with all the anomalous cases" (p.97). It therefore has the unique ability to make *general* claims about the species that has transformed not only the prospects and direction of biological life on the planet, but the geological realities, the very rocks, themselves. And yet, as Graeber continues, the discipline "resolutely refuses to do so" (Ibid.) There are many reasons this is the case, most of which are perfectly reasonable responses to its own past: from its alliance with colonialism to its positivist reinterpretation of cultures. However, in an age whose literal definition is being measured by that central disciplinary figure, anthropology must come to the table to say its piece so as to avoid the recapitulation of ethnocentrism it has long rallied against. Internally it has certainly already more than begun. Anthropologists were among the first to endeavour to build a new vocabulary, what they call a "lexicon for an Anthropocene yet unseen" (Howe and Pandian 2016), in the attempt to get a linguistic and conceptual handle on our present experiences, and those to come. However the debate concerning the placement of the geological "golden spike," the agreed-upon reference point for stratigraphic sectioning, is settled will be a reflection and designation of our very concept of the nature of our species. It is here that anthropologists must engage more fully *externally*.

The very term *Anthropocene* implies a Universal Humanity. This ontological concept, which is so dominantly embraced today, can be counted as an accomplishment of the Enlightenment. The humanist zeal for equality (in theory, certainly not in practice) expounded by that era can count as one of its children the very discipline of anthropology which from its inception was organized around the ontological unit of the "Human." Although this notion has been problematized and critiqued throughout the history of the discipline, to the point where the claims of *any* human universals were balked at as naïve, it is beyond doubt that this concept is again rearing its globe-striding head, its spectre now haunting the planet as a whole. Just as we have been implored to see the entirety of the planet as a single organism of Gaia-like qualities, we are being entreated again to see our species as a unit. Unfortunately facing Gaia as a species-whole has not functioned to induce at the very least, upon turning back to face ourselves in Gaia's gaze, the inclusion of the varied knowledge, practices, and orientations of that species in its full-fledged variation. As Murray Bookchin underlined in his book *Remaking Society* (1989), this mythical species-unit has obscured our view of ourselves, replacing any analysis of the particularities of societies, their structures, and their respective consequences, both social and ecological.

If we are not to relegate our own species to the (future) fossil record, we must be willing to operationalize everything the discipline has gleaned in its short time as a tradition and methodology. This necessitates taking the task of translation more seriously. If even to be accomplished partially (Strathern 2004), this undertaking is no small feat. Anthropology has taken great steps in this direction, following McLean's (2009) disciplinary entreatment to "be willing to engage its informants as fully fledged intellectual interlocutors and potential co-producers and to enter into dialogue with, rather than seeking to explain away, the ontological and metaphysical claims that they put forward" (p.235). However, all too often, when it comes to attempts to holistically "include" ways of knowing that differ from standard Western folk metaphysics and epistemology, especially in regards to indigenous peoples, in our institutions and governments, it comes up against Povinelli's (1995) description of the state apparatus as being the "cultural organization of Western disbelief" (p.506). Although as Linda Tuhiwai Smith wrote in *Decolonizing Methodologies* (2012), "many indigenous writers would nominate anthropology as representative of all that is truly bad about research" (p.11), we cannot abandon the project of the discipline. However, in doing so we must heed the calls to decolonize the discipline, as well as the Anthropocene itself (Davis and Todd 2017). What this might additionally do is unearth the realization that anthropology actually offers the opposite of a Universal Humanity.

As Quilley (this volume, p.217) underlines in the language of anthropology's sister discipline, the radically variable patterns of socialization among human beings in relation to their environment "give rise to very distinctive patterns of personality formation or what Elias called 'psychic habitus.'" To extend this thought, the ethnographic record contains vastly differing notions of not only human liberty individually and societally, but additionally and crucially varying conceptions of that liberty in relation to the more-than-human world in which it is enmeshed.

The ascendance of an era

Humanity's entanglement in the larger "meshwork" (Ingold 2015) of life is now undeniable, no longer made invisible by Modernist cleansings (Latour 2000). "Thinking the human at Earth magnitude" as Morton (2016, p.35) describes has made the magnitude of that earth clearly essential, not only to research and theory but to survival. The irony of course is that this transformation in the biosphere was only accomplishable in alliance with many creatures and critters privileged over others. We have never been alone, or have done anything, or will do anything, segregated, or without coexistence (Haraway 2015). However we must resist the urge to overemphasize this fact in our analysis of the generative causes, for however much we might want to address our work to these plants and animals, microbes and fungi, the multispecies assemblages, and so forth, this is, after all, and we must reckon frankly with it, a system of actions for which only human beings (at least some of them) can take responsibility.

There has thusly been a double movement towards decentring humanity as the locus of analysis in social studies and the humanities, even, perhaps a tad contradictorily, in anthropology itself, while at the same time living in an era utterly dominated by that very species. The burgeoning insistence on agency beyond the human, and the search for ways to be both free and related to that agency has turned many eyes towards if not anthropology, at least what would in past times be considered its "record." This has expressed itself as a shift in focus towards "other" cosmologies, non-Modernist modes of construing the world, a pivot to those who have not produced the Anthropocene. In the world of theory this largely has taken place under the banner of the ontological turn and post- and de-colonial thought, however much at times these two intellectual trends have overlooked some of the insights of the other. What it has forced us to reckon with, in taking indigenous ontologies as the statements of fact they are intended to be, is a view of the more-than-human world as it is found factually populated with untold agencies that one must live with in relation.

Ludwig Wittgenstein (1967) famously made the point in the *Philosophical Investigations* that if a lion could speak we would not be able to understand him for his *form of life* was too dissimilar to our own, his context too foreign. However, if we are to take the ontological turn seriously, then for instance from the perspective of those Australian aboriginals with whom Povinelli (1995) studied for whom "objects" as small as rocks *listen*, then the difference between humans, lions, and rocks is much smaller in its gulf of communication than that between those who can see that and those who cannot. Those who cannot have been taking liberties with those who can (human and more-than) since the inception of the Civilicene. It is the very distinction on which the era turns. Certain humans have construed their selves and choices to be worthy of the constriction of those of others. If humans and lions (though rocks are certainly not unaffected) are to live at all, it is this that must be undone.

Sociogenic vs. anthropogenic

The difference is clear: not all human beings have participated, or are participating, in the patterns of living that have more-than-begun to transform the global climate and biosphere. Prior to contact with European societies, the Inuit were not an "emitter nation," though they are indeed those on the front lines of that change. Satterthwaite (2009) underlines this fact, concluding that up to one-sixth of the human population "best not be included in allocations of responsibility for GHG emissions" (2009). Bruce Jennings (2018) agrees, underlining that this is a fundamental question of justice in the Anthropocene. Indeed, many indigenous peoples are, and inherently have always been, at the frontline of resistance to those changes, as they are at this very moment of writing. Would not their inclusion in the narrative of the genesis of the Anthropocene be just yet one more example of colonization and inculcation in a grand narrative not their own, a form of naturalization of cultural hegemony? This is certainly not to say that indigenous peoples are not engaged in modernity (whether by choice or otherwise), or in some capacities complicit in the destruction of land, both now and in the past. However, the marshalling of their stories and histories into the grand narrative of a Universal Humanity at fault for the Anthropocene is *historiography*. Although I certainly agree with Davis and Todd (2017) that the colonization of the Americas marked the "arrival" of the Anthropocene to these shores, and that the very act of that chronological demarcation is itself a political act, I disagree that it is a sufficient starting date. We must follow the Anthropocene back across the Atlantic to its genesis in place and time. In this way it can be seen that it is not humanity per se that has created the Anthropocene, but a specific type or subset of human beings, however large in number that group may be. As Bookchin (1989) writes "the divisions between society and nature have their deepest roots in the divisions within the social realm ... that are often obscured by our broad use of the word "humanity" (p.32). Indeed, what is being precluded by our internal conflicts is the very possibility of multi-species sociality.

In a prophetic paper entitled "Should Trees Have Standing?" Christopher Stone (1972) wrote that "perhaps someday all mankind shall be, for some purposes, one jurally recognized 'natural object.'" (p.457 – Footnote 26). In light of the fact that myriad peoples the world over have not only *not* contributed to ushering in the "Anthropocene," but have actively resisted it, it would be unjust to include them in any legal "natural (hyper)object" whose agency resulted in our current wicked dilemma. It is precisely this tension between the "liberated" humanity and its incompatibility with a finite planet (or a species to which it is worth relating) that marks the Civilicene. As Jennings (2018) underlines, this is not true freedom, writing that being "at liberty to behave in ways that are ecologically irresponsible and destructive is not to be liberated; it is to be dominated by technology and desire" (p. 88). This is to say nothing of the liberty (or lack thereof) of the more-than-human world.

Malm and Hornborg (2014) make the point quite clearly, that "[r]ealising that climate change is 'anthropogenic' is really to appreciate that it is *sociogenic*" (p.66). This brings into question the entirety of the Anthropocene narrative, and has crucial implications for the contested debate surrounding the dating of its inception. The question becomes, in what kind of *society* is the *genesis* found? Scholars such as Jason Moore have suggested more specifically sociogenic terms like "Capitolocene" (2015). However, as Haraway readily declares, capitalism "is a late development!" (Haraway et al. 2016, p.556) and therefore our predicament requires an even newer word to signal its primary character. One proposed starting date espoused by those like scientist Cesare Emiliani (1993) is the addition of 10,000 years to our present calendar to mark the Human Age, the beginning of our "species" era being set at the Neolithic Revolution. Although the scale of change would be unremarkable today, this marked a qualitative shift in the relationship of certain human beings to their environment, which over time resulted in the radically quantitative change we see now. This approach has been labelled the "early Anthropocene" hypothesis (Malm and Hornborg 2014, p.63) or what Scott (2017) has called the "thick" Anthropocene, rather than the "thin" one, which he traces back to the mastery of fire by *Homo erectus*. Contained implicitly within the framing of these arguments as the "early Anthropocene" is its own terminological downfall. If, as Morton (2016) writes, "[t]he ecological emergency we call the present ... has been happening in various forms for twelve thousand years" (p.150), this puts into clear relief the fact that this is merely *one strand* of the human story. That exact point traditionally taught in high school classes demarcating the boundary between the categories of pre-history and history can be thought of not as the beginning of our species' history, but as the point of departure for our present catastrophe.

The very question *of history* is one and the same as that of the Anthropocene. The historiographical construction of identity between the inception of the written word and any events worth remembering is of course a civilized form of ethnocentrism which devalues all other forms of past-reckoning and knowing, whether oral, material, genealogical, or placed (Smith 2012). The contestation of that history is ongoing and crucial as it is daily carried out in courts and communities the world over, where indigenous people and others attempt to reclaim and participate in the construction of the narrative of their own lands, their own pasts, and peoplehood. As Smith (2012) writes, history "is the story of the powerful and how they became powerful, and then how they use their power to keep them in positions in which they can continue to dominate others" (p.35). In actuality, it might be said that the oral histories and genealogies of indigenous peoples the world over provide the only actual truly long-term history we have. Smith continues that "[t]o hold alternative histories is to hold alternative knowledges. The pedagogical implication of this access to alternative knowledges is that they can form the basis of alternative ways of doing things" (p.36).

The Civilicene

The proposed term that might be closest to my appraisal is what Haraway and others have termed the Plantationcene (Haraway et al. 2016). Haraway (2015) writes elsewhere that this term was generated to describe "the devastating transformation of diverse kinds of human-tended farms, pastures, and forests into extractive and enclosed plantations, relying on slave labor and other forms of exploited, alienated, and usually spatially transported labor" (p.162 – Footnote 5). As she said of the Capitalocene, I would argue that even this is a late development. So what is this qualitative difference that puts some humans, or rather cultures, on a species-on collision with the systems of life that underlie them, and some do not? The present era might be said to be the result of taking certain *liberties* with the *agency* of others. This shift, which might be described (however difficultly) as from trust to domination as anthropologist Tim Ingold (2000) termed it, from relying on the "giving environment" (Bird-David 1990) to forcing the environment to give, amounting to what Quilley has called in this volume disenchantment and disembedding, started a revolution that continues in the Amazon today, as rainforests fall to the onslaught of soybeans and cattle. As anthropologist Noboru Ishikawa simply put it, "to me, plantations are just the slavery of plants" (Haraway et al. 2016, p.556). The archaeobiological evidence of the extent of this slavery (which was not in any way constrained only to plants), as Smith and Zeder (2013) indicate, could easily mark the Anthropocene. Although the gradation between these two styles of living are messy and exist on a spectrum (see, for instance, Leach 2003) there is still something particular about cultures that live by what Morton (2016) calls "agro-logistics" and why he refers to we of the global agricultural civilization as "Mesopotamians," those who continue the revolution started in that ancient land. Tuck and Yang (2012) point out that the settler's very identity was (and is) constituted "by making the land produce, and produce excessively, because 'civilization' is defined as production in excess of the 'natural' world (i.e. in excess of the sustainable production already present in the Indigenous world)" (p.6).

I need not restate in detail the *still* dominant narrative of our culture, that Civilization (with all its faults) *freed* us from the toil of existence in the "wild." That the state it created insulated us from the nature whose teeth and claws would otherwise leave us red. The complex stratified societies that it birthed divided labour so as to give our species the liberty to devote itself to the whole host of high culture and science of which we are heirs. In modern times it freed us from the shackles of convention, religion, kinship, and so on. This is of course comforting historiography, though not incorrect in its facts, but grossly misleading in its omissions. This all came at a cost, and not only the cost described by those like Jared Diamond (1999) in terms of our health. It freed us from our own liberty of movement, constrained our agency at the hands of the states, and shackled us to the tending of our domesticated foods and the sedentary and dangerous lives that entailed. This is what Jennings (2018) described, drawing on philosopher C.B.Macpherson, as the "surrendering of the developmental power of the self to the extractive power of others" (p.90). Not only is this true for inter-human relations, but it constituted the ever-expanding enslavement and instrumentalization of non-humans that continues to this day. This cost or "externality" comes now too in the form of carbon dioxide. Although the gathering into the city, the *civitas*, in some form likely predates the state and agricultural civilization as Scott (2017) has summarized, I continue to use the term *civilization* so as not to avoid, but to fully embrace, the deeply charged history of the term and its sociogenic consequences.

Alternatives have existed continuously since the Neolithic inception of an extractive agricultural society based upon command and control of non-human nature. Indeed, concurrent with this revolution were myriad societies throughout the globe that did not, and at times actively chose not to engage in this project of Civilization. This is certainly not new subject matter for the discipline seeing as the archaeological record is literally littered with societies that have faced analogous situations to ours. Evidence of both mysterious collapse as well as conscious abandonment abound (Tainter 1988; Scott 2017). It is clear that the lessons of our ancestors are available to us. Not only their failures, but their successes, past and present, too. The Inuit, or the Ju/'hoansi, or any non-agricultural people were not left behind, are not *pre*-agricultural or *pre*-civilized, but operating, as richly and with the same full expression of humanity, entirely separate from the entire game of agro-logistics.

Definitional intervention

An elephant (or *anthropos* rather) in the room that must be acknowledged is the whole strain of writers and thinkers outside the academy proper, from John Zerzan (1994) to Derrick Jensen (2006), who draw on anthropological and other disciplinary work, arguing in many ways in a similar vein. For instance, Zerzan, who asserts that civilization itself is not only unsustainable but undesirable, claiming that the only viable future would be, in his words, "primitive," claims kinship with Morton who underlines his divergence (see 2015). Their work and its tendency towards primitivism, as well as the widespread misuse and misinterpretation, often for political and ideological reasons, of anthropological data and insight in general, is likely one of the reasons for the discipline's misgivings (sometimes with good reason) about entering fully this discussion. Regardless of normative claims, it is more than possible that the development of civilization was inevitable. It was indeed invented independently in multiple locations throughout the world and there are those who argue that it is an inescapable response to the stable climactic period of the Holocene (Bennett 2017), or that the Anthropocene and the Holocene are coextensive (Smith and Zeder 2013). I do not wish to make statements on the above claims, but merely to make clear that regardless of whether civilization is redeemable, or desirable, it has happened, and it is the genesis of our current moment. The path-dependencies of what Smith and Zeder (2013) have called our "major new type of human niche construction" (p.9) inaugurated in the Neolithic have led to the one degree of warming we are *already* experiencing globally.

I indulge in this definitional discussion because, I think, echoing anthropologist Marilyn Strathern (1992), that "it matters what ideas we use to think other ideas with" (p.10). The distinction between species and society is crucial. As Haraway (2016) notes, if indeed the Anthropocene reveals our present situation as an inevitable result of the human character, unfortunately stories like that end badly. The evidence of this kind of equivocation is everywhere. A recent piece of telling evidence is the title of a *New York Times* article detailing the conclusions of the IPBES Global Assessment Summary for Policymakers that biodiversity loss has reached an unprecedented level, endangering the future of civilization. The original title cited civilization itself as the culprit, only to be changed within a few hours to cite human beings as the perpetrators (Plumer 2019). This kind of misanthropy, even of the environmental kind, has been roundly criticized by Bookchin (1989). He would agree that what settling this definitional ambiguity does is open a space in which we can think that it was not the entering of the human onto the world stage that started this path, but a new kind of society–nature relation that developed in a particular time and place, and operates in a particular way.

This does not necessitate a specific response such as a "return" to a prelapsarian state. Though Van den Burgh (2011) is perhaps correct in *practical* terms in writing that, in regards to contemporary calls for degrowth, it "is unlikely that hunter-gatherers or Henri David Thoreau ("Walden") can serve as a role model for them" (p.883), it certainly depends on what a role model might mean. Although this chapter is but a cursory sketch of what this reframing could do for our thinking and praxis, there is promising work in this area already. The "decelerationist" pathways as possible future directions for our global society sketched out by Hensher in this volume is a good example, which itself draws upon the work being done in degrowth scholarship. Some researchers in this area are precisely looking to "non-civilized" cultures as stimulus for the sketching of a future steady-state economy (see Kallis et al. 2018, for example). It is quite possible, as David Graeber (2016) has written, that our technological developments hold out the possibility of *more* potential social and ecological relations, rather than fewer. That is, in learning from, as I have argued, other ways we have been (and are) human, we might use our current technological complexity to live otherwise, not in old, but in utterly new ways.

The descent of an era

Perhaps in an age so young the idea of the golden spike is not yet relevant to our own self-conception. Rather, I think we should, and mostly have, been thinking of the Anthropocene from an historical standpoint rather than a geological one, as Jason Moore clearly distinguishes (2015). This is surely an explanation for the quite understandable disciplinary reticence of geologists, and the disbelief of that reticence by those doing thinking in the Arts. Haraway (2015) has written that it might be better thought of in terms of being a boundary event (geologically), rather than an epoch. Regardless, it can easily be described as one predicated upon imaginary boundaries (Latour 2000; Haraway 2008) between humans and nature, and between civilization and any alternative. The rocks may not demarcate the present human-created era to the satisfaction of geologists, but they may very well be listening.

If we are to usher in a second era in which the human has a majority impact, whether that be described as the "Chthulucene" (Haraway 2015), or the "Ecozoic" (Swimme and Berry 1992), in which our dealings with all the beings of this earth are mutually enhancing, we must learn not only to live in relation to the agency of non-human beings, but thusly to constrain our own in relation to them. In an effort to describe the kind of more-than-human oriented conception of liberty appropriate to that future world, Jennings (2018) develops a notion of "relational liberty" where interdependence is constitutive of, and lays the very grounds for, the possibility of freedom itself. In many ways this holds similarities with much well-familiar anthropological work on societies whose conception of the self is in myriad nuanced ways, constituted not by its singular identity but its relations. Unsurprisingly, these societies are generally those non-civilized ones of which we have been speaking. However, this kind of relationality can never be thought of as settled or uncomplicated. It requires constant navigation and development, and is rife with inherent conflict, not the least of which is that we live upon the deaths of others. This accounts, for instance, for the complex procedures of distancing, seduction, rapport, and trickery that characterize the cosmologies of hunting societies, which must constantly negotiate their relationship with their prey (and their respective animal masters) (see, for instance, Willerslev 2007). For animistic societies like these, what Jennings sets out as the necessary basis for true liberty in the Anthropocene, bringing the more-than-human world into the arena of moral participation, characterized by membership, recognition, and solidarity, has already long been accomplished. These societies have developed more complex cultural technologies for this kind of relational participation than those often put forward in the realms of contemporary political philosophy. If we are unable to overcome not only our species, but our cultural narcissism, and understand anthropology's revelation that we can *be* other kinds of people, we will forever remain a lonely species, both ontologically, and, regrettably, existentially. We will live not in the Ecozoic, but be banished to what E.O. called "the Eremocine, the age of loneliness" (Jarvis 2018).

For all of our adeptness as a species it is quite clear that there is still much we do not know about ourselves. Though our struggle for self-understanding is nothing new, if we were to cease to exist because of its lack, it would not only be an occupational hazard for me as an anthropologist, but a terrible loss for the multi-species assemblage that is the earth. And yet, the encouraging fact is that we *do know* that we have, and do, live in ways in which all members of the ecology are free, and also not destined for climate catastrophe. As Bennett (2017) summed up, "we need new ancestors. The old ones have been exhausted and are exhausting us." This, in the end, is a plea for my native discipline to step up to the plate (that literally may have its name stamped upon it). Not only do the universities (Leader 2017) need anthropology now more than ever, we all do. We need a truly *activist anthropology*, one that marshals its intellectual heritage, its cultural insight, and its powerful lens and theory to grapple with the largest struggle this species has ever faced. We can use all the insight we can get, as the scale of change required to transition out of the Civilicene is, as Haberl et al. (2011) describe, comparable to that from hunter-gatherer to agrarian societies and from those to the industrial societies of the present. Likely it is larger still. If we do not want to join that shelf of collapse literature for a possible future society's (or species') perusal, we must be wary, for these kinds of transitions have often marked that tipping point for complex societies.

Conclusion

To avoid our time being the "last biodiverse geological epoch that includes our species too," (Haraway 2016, p.37) and to aim to limit our destruction to the shortest possible length, we must rewrite our species narrative to allow us to *tell different stories about ourselves*, ones where the earth is no mere backdrop, but the full cast of characters. If we're going to tell a Big Story like the Anthropocene, we might as well be specific. And this is the task to which anthropology has always been destined. Although as Malm and Hornborg (2014) wrote "'the Anthropocene' is already an entrenched concept and mode of thinking" (p.66) – that is, at least outside of geology, I do not think the matter is sufficiently settled. From a multi-species perspective, we can certainly see how, as they continue, it "might be a useful concept and narrative for polar bears and amphibians and birds who want to know what species is wreaking such havoc on their habitats" (p.67). However, even here I do not think the concept fulfils its role, as polar bears and amphibians and birds are more than anyone cognizant of the differing relations they have with human beings. Let us not overlook the fact that these beings are relational and conduct these relations with humans in a variety of ways throughout the world, some of which are mutually enhancing. What the useful ascendance of the concept of the Anthropocene *does* do, is give us the opportunity to right the narrative, to put front and centre the facts of the matter, and to hopefully then overcome our teleology.

It is quite possible that our civilization will collapse, or at least will undergo "involuntary degrowth" as Hensher cites in this volume. If and when this happens, there is no doubt that its effects on the planet will long outlast it. The question then becomes: will we as a species, and in what capacity? Our self-conception is defined by the stories we tell about ourselves. What anthropology brings to the table is that it might teach us that we are *at liberty* to do otherwise, and that we *have the agency* to change our ways. It matters what stories tell stories. And it matters what boundaries make boundaries.

References

Bennett, E. (2017). "Seeds of a good anthropocene." *The Long Now Foundation*. Retrieved from longnow.org/seminars/02017/nov/20/seeds-good-anthropocene/
Bird-David, N. (1990). "The giving environment: Another perspective on the economic system of gatherer-hunters." *Current Anthropology*, *31*(2): 189–196.

Bookchin, M. (1989). *Remaking Society*. Montréal: Black Rose Books.

Davis, W. (2009). *The Wayfinders: Why Ancient Wisdom Matters in the Modern World.* Toronto: House of Anansi.

Davis, H., and Todd, Z. (2017). "On the importance of a date, or decolonizing the anthropocene." *ACME: An International Journal for Critical Geographies, 16*(4): 761–780.

Diamond, J. (1999). "The worst mistake in the history of the human race." *Discover Magazine.* Retrieved from http://discovermagazine.com/1987/may/02-the-worst-mistake-in-the-history-of-the-human-race

Emiliani, C. (1993). "Correspondence – Calendar Reform". *Nature, 366*: 716.

Gibson-Graham, J. K. (2008). "Diverse economies: Performative practices for other worlds." *Progress in Human Geography, 32*(5): 613–632.

Graeber, D. (2004). *Fragments of an Anarchist Anthropology*. Chicago: Prickly Paradigm Press.

Graeber, D. (2016). *The Utopia of Rules: On Technology, Stupidity, and the Secret Joys of Bureaucracy.* Brooklyn, NY: Melville House.

Haberl, H., Fischer-Kowalski, M., Krausmann, F., Martinez-Alier, J., & Winiwarter, V. (2011). "A socio-metabolic transition towards sustainability? Challenges for another Great Transformation." *Sustainable Development, 19*(1): 1–14.

Haraway, D. (2008). *When Species Meet*. Minneapolis: University of Minnesota.

Haraway, D. (2015). "Anthropocene, Capitalocene, Plantationocene, Cthulucene: Making Kin." *Environmental Humanities, 6*: 159–165.

Haraway, D. (2016). *Staying with the Trouble: Making Kin in the Chthulucene.* Durham: Duke University Press.

Haraway, D., Noboru Ishikawa, S. F. Gilbert, K. O., Tsing, A. L., & Bubandt, N. (2016). "Anthropologists Are Talking – About the Anthropocene." *Ethnos, 81*(3): 535–564.

Howe, C., & Pandian, A. (2016). "Lexicon for an anthropocene yet unseen." *Theorizing the Contemporary, Cultural Anthropology.* Retrieved from https://culanth.org/fieldsights/803-lexicon-for-an-anthropocene-yet-unseen

Ingold, T. (2000). *The Perception of the Environment: Essays in Livelihood, Dwelling and Skill.* London: Routledge.

Ingold, T. (2015). *The Life of Lines*. London: Routledge.

Jarvis, B. (2018). "The insect apocalypse is here." *The New York Times.* Retrieved from https://www.nytimes.com/2018/11/27/magazine/insect-apocalypse.html

Jennings, B. (2018). "Liberty: The future of freedom on a resilient planet." In D. A. DellaSala, & M. I. Goldstein (eds.), *The Encyclopedia of the Anthropocene*, Oxford: Elsevier. 4, 87–94.

Jensen, D. (2006). *Endgame*. New York: Seven Stories Press.

Kallis, G., Kostakis, V., Lange, S., Muraca, B., Paulson, S., & Schmelzer, M. (2018). "Research on degrowth." *Annual Review of Environment and Resources, 43*:291–316.

Latour, B. (2000). *We Have Never Been Modern*. Harlow, Essex: Pearson Education.

Leach, H. (2003). "Human domestication reconsidered." *Current Anthropology, 44*(3); 349–368.

Leader, G. (2017). "Universities need anthropology now, more than ever." *Huffington Post: American Anthropological Association.* Retrieved from https://www.huffingtonpost.com/american-anthropological-association/universities-need-anthrop_b_1257 6982.html?ec_carp=3070557437376468536

Malm, A., & Hornborg, A. (2014). "The geology of mankind? A critique of the anthropocene narrative." *The Anthropocene Review, 1*(1): 62–69.

McLean, S. (2009). "Stories and cosmogonies: Imagining creativity beyond 'nature' and 'culture'." *Cultural Anthropology, 24*(2): 213–245.

Moore, J. (2015). *Capitalism in the Web of Life. Ecology and the Accumulation of Capital.* London: Verso.

Morton, T. (2015). "What is dark ecology?" *Changing Weathers.* Retrieved from www.changingweathers.net/en/episodes/48/what-is-dark-ecology

Morton, T. (2016). *Dark Ecology: For a Logic of Future Coexistence.* Wellek Library Lectures in Critical Theory. New York: Columbia University Press.

Plumer, B. (2019). "Humans Are Speeding Extinction and Altering the Natural World at an 'Unprecedented' Pace." *The New York Times.* Retrieved from https://www.nytimes.com/2019/05/06/climate/biodiversity-extinction-united-nations.html?smid=fb-nytimes&smtyp=cur&fbclid=IwAR2IjCRglG93Kpj5Jo0tjySKnammEV6j8BsCe8d-DoYENnTufUvXtNruccE

Povinelli, E. (1995). "Do rocks listen? The cultural politics of apprehending Australian aboriginal labor." *American Anthropologist. New Series,* 97(3): 505–518.

Satterthwaite, D. (2009). "The implications of population growth and urbanization for climate change." *Environment & Urbanization,* 21: 545–567.

Scott, J. (2017). *Against the Grain.* New Haven: Yale University Press.

Smith B. D., & Zeder, M. A. (2013). "The onset of the anthropocene." *Anthropocene.* Retrieved from http://dx.doi.org/10.1016/j.ancene.2013.05.001.

Smith, L. (2012). *Decolonizing Methodologies: Research and Indigenous Peoples* (2nd ed.). London: Zed Books.

Stone, C. D. (1972). "Should trees have standing?–Towards legal rights for natural objects." *Southern California Law Review,* 45: 450–501.

Strathern, M. (1992). *Reproducing the Future: Essays on Anthropology, Kinship, and the New Reproductive Technologies.* New York: Routledge.

Strathern, M. (2004). *Partial Connections (Updated ed.).* Walnut Creek, CA: AltaMira Press.

Swimme, B., & Berry, T. (1992). *The Universe Story. From the Primordial Flaring Forth to the Ecozoic Era–A Celebration of the Unfolding of the Cosmos.* San Francisco, CA: Harper San Francisco.

Tainter, J. (1988). *The Collapse of Complex Societies.* Cambridge: Cambridge University Press.

Tuck, E., & Yang, W. (2012). "Decolonization is not a metaphor." *Decolonization: Indigeneity, Education and Society,* 1(1): 1–40.

Van den Bergh, J. C. (2011). "Environment versus growth—A criticism of "Degrowth' and a plea for 'A-Growth'." *Ecological Economics,* 70(5): 881–890.

Willerslev, R. (2007). *Soul Hunters: Hunting, Animism, and Personhood among the Siberian Yukaghirs.* Berkeley: University of California.

Wittgenstein, L. (1967). *Philosophical Investigations.* Oxford: Basil Blackwell.

Zerzan, J. (1994). *Future Primitive: And Other Essays.* Brooklyn, NY: Autonomedia.

13

DEFENDING AND DRIVING THE CLIMATE MOVEMENT BY REDEFINING FREEDOM

Aaron Karp

Climate change as an issue of economic and cultural transformation

Climate change is mainly understood by the public and by activists as an *energy problem* that can be solved through a rapid transition from fossil fuels to renewables. What is less appreciated is that climate change is also an *economic problem* requiring a simultaneous transformation of the present economy. If we assume that currently non-existent carbon-negative technologies do not come into being (Fuss et al., 2014), then in order to hold to 2°C – the only warming limit consistently identified in international climate discussions – rich nations must reduce emissions at rates above 10% per year, a feat that has never been achieved (Anderson, 2015). For context, emissions reductions greater than 1% per year have historically occurred in situations of economic upheaval or recession (Stern, 2007).

To overcome the apparent link between economic turmoil and serious climate action we can look to the discipline of ecological economics, which outlines the policies and institutional changes that could transform the current profit-driven economy into a non-growing, steady state economy (SSE) that has as its goal the meeting of basic human needs within ecological limits. Presently, if fossil-fuelled consumption levels do not remain high enough, then economic growth reverses into recession. This outcome would likely make rapid emissions reductions socially and politically untenable. Establishing a SSE would eliminate the unstable grow-or-contract nature of the current economy and could allow for a swift transition to renewables while avoiding economic breakdown.

Because climate change is an *economic problem* requiring an economic transformation, it is also a *cultural problem* requiring a cultural transformation. The current economy allows individuals to consume as much as they can afford, and consumption plays an outsized role in our understanding of freedom. A SSE preserves vital natural systems by establishing limits to consumption, a fact that elites who own and manage the economy will seek to exploit. Why? Growth is treated as the best (and only) way to improve the economic situation of working-class citizens – a substitute for equality – and every step towards a SSE would spotlight the need to redistribute wealth. A non-growing economy requires clearly defined limits to economic inequality and calls into question the very existence of profit-maximizing institutions and the exorbitant private fortunes that exist today. To combat this threat to their financial interests and dominant social position, elites will vigorously oppose this transition. A key strategy will involve attempts to generate public opposition by arguing that to limit consumption is to undermine an essential freedom. Establishing a SSE will thus only be possible if a new cultural understanding of freedom gains legitimacy over the consumerist definition, and climate activists must lead that campaign.

"The contest for legitimacy is a public battle for the supremacy of particular frames that underpin the legitimacy of specific norms and of the organizations and institutions that promulgate them," writes Julie Ayling (2017, pp. 362–363). Activists fight this battle by shifting discourse and the public's understanding of core cultural ideas. Though industry possesses significant economic and political advantages, activist groups "typically do enjoy considerable 'discursive' and 'symbolic' power, meaning battles over *ideas* and *legitimacy* tend to be less one-sided," observes Fergus Green (2018, p. 109). By generating a society-wide discussion aimed at redefining the concept of freedom, climate activists can protect their movement. A SSE embodies certain principles: the importance of limits, the equality of human beings, the ethic of sufficiency, and others. By asserting a new understanding of freedom that features these ideas, activists will undercut elites' attempts to delegitimize the movement through appeals to the consumerist definition of freedom. This redefinition process can also drive the movement towards a SSE. The promise of freedom has historically been a central motivation of social movements, and by asserting a new, inspiring vision of eco-democratic freedom—and the climate movement as a vehicle for that vision—activists gain a potentially significant source of engagement.

The next section reviews the early history of corporate elites' crusade to define freedom as consumption. Educating the public about this history is an essential part of redefining freedom, as it delegitimizes the corporate definition and shows that consumerist lifestyles had to be forced upon the public. The following section outlines an ecological and democratic understanding of freedom by sampling the views of prominent classical liberals and ecological thinkers. The final sections examine the reasons why activists must launch a mass-communication campaign to assert the eco-democratic definition over the consumerist definition.

Consumption as freedom: Business shapes culture in its preferred image

When one considers the world-leading consumption levels of US citizens, it is easy to imagine that daily life in the US was always defined by consumption. But prior to the start of the 20th century, thrift had been a classic feature of American culture (Ewen, 1976). This trait became a major problem for business with the development of mass production, which for the first time led to a significant surplus of consumer goods beyond what citizens required to meet their basic needs. Business leaders feared a permanent crisis of overproduction, with social historian Stuart Ewen noting that consumerism "emerged in the 1920s not as a smooth progression from earlier and less 'developed' patterns of consumption, but rather as an aggressive device of corporate survival" (1976, p. 54). Business thus became preoccupied with the challenge of turning the American cultural ethic of sufficiency into one of constant consumption.

However, "underconsumption" was not the only crisis facing corporate elites. Around 1900, mass media that could bring news and other information to communities across the US were just getting established (Ewen, 2003). This far-reaching press was informing the public about the increasing control of social conditions by large corporations and the violence unleashed against workers attempting to organize and improve their conditions (Ewen, 2003). An increasingly politically powerful public was forming solidly anti-corporate sentiments. Business had tried to impose industrial discipline on American workers through horrific violence for decades, but began to shift towards organized propaganda, harnessing the new channels of communication to re-establish its social legitimacy. The creation of a consumer culture, it was thought, could address both problems.

The power of propaganda was demonstrated by President Woodrow Wilson's Committee on Public Information during World War I, which successfully transformed a pacifist population into one clamouring for war (Ewen, 2003). Edward Bernays, a member of the Committee and later the recognized "father of the public relations industry," brought the tested techniques of manipulation to the private sector. He observed that

> mass production is profitable only if its rhythm can be maintained—that is, if it can continue to sell its product in steady or increasing quantity … today supply must actively seek to create its corresponding demand … [and] cannot afford to wait until the public asks for its product; it must maintain constant touch, through advertising and propaganda … to assure itself the continuous demand which alone will make its costly plant profitable.
>
> *(Bernays and Miller, 2005, p. 84)*

Both public relations and advertising would develop into their own sectors of business in the 1920s.

Control would be gained by associating individual liberty with the purchase of goods (and the corporations producing them) in the public mind. In 1924, retail magnate Edward Filene observed that

> modern workmen have learned their habits of consumption and their habits of spending (thrift) in the school of fatigue, in a time when high prices and relatively low wages have made it necessary to spend all the energies of the body and mind in providing food, clothing and shelter. We have

no right to be overcritical of the way they spend a new freedom or a new prosperity until they have had as long a training in the school of freedom.

(as cited in Ewen, 1976, pp. 29–30)

An expanding corporate propaganda machine would provide that training. "During the 1920s," Ewen observes, "advertising grew to the dimensions of a major industry" (1976, p. 32). Between 1900 and 1930, national advertising revenues grew from $200 million to $2.6 billion, a thirteen-fold increase (p. 62). Growing investment produced great successes, with business increasingly associating itself with liberty. Historian Kerryn Higgs (2016) notes how marketing publication "*Advertising Age* credited the National Chamber of Commerce with divorcing the word 'big' from the word 'business' in the public mind" (p. 177). "Private enterprise" was replaced by "free enterprise," and Henry Link of the polling firm Psychological Corporation later promoted "a transfer in emphasis from free enterprise to the freedom of all individuals under free enterprise; from capitalism to a much broader concept: Americanism." Link recognized that, unlike "free enterprise," "Americanism" possessed a "terrific emotional impact" (p. 179).

The reach of this propaganda machine was exemplified by the massive campaign launched by the National Association of Manufacturers in the 1930s to define the "American Way of Life." It replicated the World War I model, establishing local Committees on Public Information composed of influential community leaders throughout the country. These agents "funneled articles, features, and films to newspapers, radio stations, and movie theaters," sent speakers to "every local group of any sort," and "distributed pamphlets and weekly bulletins to schools, clubs, and libraries" (Higgs, 2016, p. 175). Particular care was taken to target the young:

> Aware that the adult population was cynical about the corporate claim to 'service,' they aimed specifically at schools, where *Young America*, their weekly children's magazine that portrayed capitalism as dedicated to looking after them and their communities, was sent to thousands of teachers, who used them in classroom assignments. *You and Industry*, a series of booklets written in simple language, linked individual prosperity to unregulated industry, and was distributed to public libraries everywhere. One million booklets were distributed every two weeks by the US Chamber of Commerce, which, along with the giant industrial corporations, was also involved in the campaign.
>
> *(Higgs, 2016, p. 175)*

The corporate elites driving the expansion of a consumer culture were joined by economists calling for a "new economic gospel of consumption" (Higgs, 2016, p. 71) and political leaders offering enthusiastic support. Higgs (2016) notes that

President Herbert Hoover's 1929 Committee on Recent Economic Changes welcomed the demonstration 'on a grand scale [of] the expansibility of human wants and desires,' hailed an 'almost insatiable appetite for goods and services,' and envisaged 'a boundless field before us ... new wants that make way endlessly for newer wants, as fast as they are satisfied.'

(p. 72)

This collective effort would eventually culminate in the society that the American people (and those of other wealthy countries) know today, in which individuals see themselves as consumers rather than citizens.

An ecological and democratic understanding of freedom

A brief sampling of the history of corporate culture gives us a sense of the effort that business has put into shaping the public's understanding of freedom in its preferred image. Before we explore why activists need to stimulate a societal rethink of freedom, we must reflect upon the concept that will replace the corporate definition.

When we recognize that addressing climate change requires us to create a fundamentally different economy, we must reckon with the fact that ordinary people have little say over the structure of the economy and thus over their own fate. Elites will not willingly make these changes. In order to transform the economic system, the public must gain control of it – in this way, economic democracy is vital to achieving a sustainable society. An understanding of liberty in the age of climate crisis must therefore foreground the freedom of the public to shape the economy.

A new definition of freedom must also highlight the importance of limits. One can find strong support for limits within classical liberal thought, which today is often claimed by wealthy industrialists as a moral foundation for unimpeded, self-interested action. However, two of freedom's greatest theorists, Wilhelm von Humboldt and John Stuart Mill, recognized that limits could rightly be placed on individual liberty. Even in asserting freedom's importance for human development, Humboldt always recognized that it exists within justifiable limits. Not only must our sphere of action preserve the equal rights of others, but restrictions on our action are warranted when "freedom would destroy the very conditions without which not only freedom but even existence itself would be inconceivable" (Humboldt and Burrow, 1993, pp. 144–145). In Mill's view, a SSE represented an important step in human advancement:

It is scarcely necessary to remark that a stationary condition of capital and population implies no stationary state of human improvement. There would be as much scope as ever for all kinds of mental culture, and moral and social progress; as much room for improving the Art of Living, and much more likelihood of its being improved, when minds ceased to be engrossed by the art of getting on.

(1909, p. 751)

These perspectives remind us that limits are not only necessary to preserve freedom and even existence, but can also act as a catalyst towards forms of progress more aligned with human flourishing.

Contemporary perspectives further inform our understanding of ecological freedom. Though some academics seem to take the market-centred, acquisitive definition of freedom shaped by corporations as *the* definition (Dibley, 2012), others remind us that such ideas can be defined differently and in ways that are more descriptive of reality. Consider Peter Brown, who recognizes that "we are required to re-examine, and ultimately to redefine the emancipation project. The narratives from which we currently take our bearings are simply not true to our circumstances" (2012, p. 7). Brown calls for the recognition of human equality to temper our understanding of freedom.

> We must see that how we live is often unavoidably harmful to others. There are no actions that affect us alone ... In a world of limits, liberty may be legitimately exercised only if one is using only his/her fair share of low entropy sources and sinks ... All persons in all cultures and all generations have equal moral claims to flourishing, constrained and enhanced by the claims of other species for their place in the sun. We are not the chosen species, or the chosen people. This, if you like, is the new emancipation.
>
> *(2012, p. 14)*

The corporate messages flowing through society have conjured a myth of individualism at the heart of conceptions of liberty, which ignores the fact that society is only possible through the care we provide and work we do for one another. Bruce Jennings (2015) argues that our understanding of freedom must be informed by the relationships we have with others and with the natural world. Freedom should be recognized as a social practice arising from the bonds of interdependence that we share. This view elevates each person to the status of subject rather than object, and does not privilege the individual over the community – rather, balance is sought between the flourishing of each. It entails a recognition that we are only free to the extent that others are, too. Power relationships should be examined and the ability to make decisions should be distributed to all affected by them. These guiding principles suggest that "the message of planetary boundaries and the end of the liberal era of cheap fossil carbon is not the bad news of lost liberty but the promise of a newfound freedom—a more humanly fulfilling kind of liberty" (p. 307).

Jason Lambacher (2009) believes that activists have avoided discussions of freedom because many citizens in wealthy nations have been free to "live in ways that *appear* to ignore ecological limits as if they were not there. In fact, this is a vital issue – our dominant political concepts, such as freedom, have not yet become ecological" (p. 32). Our ways of life, and every layer of society, must be transformed around the reality of limits. Figuring out how to live sustainably will of course be an ongoing process, part of learning how to be free on a finite planet. "What is important is not the validity of a single approach to environmental sustainability," he writes, "but rather that people feel inspired by the challenge of freely creating an ecologically responsible culture" (p. 41).

An ecologically compatible freedom will foreground democracy, both political and economic. It will acknowledge that liberty – and life – cannot exist without limits. It will be shaped by our recognition of interdependence and human equality. It will balance the interests of both individual and community. It will be broader than the simple consumption of goods, a vision encompassing the many parts of human nature left behind by the current society. Activists must assert and continue to develop this eco-democratic definition of freedom, both to protect and to drive the movement towards a SSE.

Redefine freedom to protect the movement

It is clear that as activists recognize the need to establish a SSE and begin to vocalize their demands, business will leverage the vast communications system it has built and the cultural cues it has implanted to make this transformation appear catastrophic. To protect the climate movement, citizens must learn how the consumerist definition of freedom has been carefully constructed as a pillar of a culture that serves corporate interests – the result of a century of corporate PR and advertising campaigns rather than a signifier of the inherent acquisitiveness of human nature. By educating the public about this history and exposing the narrowness of the consumerist definition as compared to the eco-democratic definition, which better supports human flourishing, activists delegitimize "consumerist freedom" and undercut those corporate messages that utilize it. The following subsections illustrate why redefining freedom is an essential defensive strategy by exploring the nature of the counterattack activists will have to withstand.

The corporate culture machine

Stuart Ewen (1976) notes that it would take a few decades, but the post-World War II period finally realized elites' vision of the mass-consumption society. Advertising and public relations deserve a lot of credit for these developments. "The new society was one which distributed *culture* on a mass scale," Ewen writes. "This triumph over the locality of people's lives as a source of nurturement and information is, perhaps, the monumental achievement of twentieth-century capitalism: centralization of the social order" (pp. 206–207).

Activists must recognize the extent to which corporate elites control society's information systems and thus have the means to dominate social narratives.

A primary consideration is that major media institutions are themselves corporations whose advertising clients are also corporate entities, and all share an interest in burying or delegitimizing challenges to the profit-driven economy. Sociologists highlight additional contours of the modern propaganda machine by exploring climate science denial networks. Wealthy family foundations now funnel untraceable "dark money" to various cultural and political causes, obscuring the support provided by specific individuals and organizations (Brulle, 2014). Corporate think tanks have also proliferated, set up to constantly generate ideas and literature that can be passed off as independent research or science (Dunlap and McCright, 2011). Front groups and astroturf campaigns further obscure reality, suggesting independent or even grassroots movement support for corporate positions. An online right-wing echo-chamber regularly amplifies baseless or conspiratorial stories through social media networks until they reach mass media channels (Dunlap and McCright, 2015). These are some of the narrative-shaping realities with which activists must contend.

The public's vulnerability

Another component of the threat facing the climate movement is the public's vulnerability to manipulation through claims of "curtailed consumerist freedom." Developing an ecological consciousness within a consumer society presents significant challenges (Hamilton, 2010; Solomon, Greenberg, and Pyszczynski, 2004). Clive Hamilton (2010) warns that "Consumption behaviour and the sense of personal identity are now so closely related that a challenge to someone's consumption behaviour may be a challenge to their sense of self" (p. 574). This identity-linked dependence on consumption can be exploited by corporate elites and mobilized against the transition to a sustainable society. In Hamilton's view, consumer identity can only change with a massive environmental calamity or a widespread loss of confidence in consumer life, thus the task of achieving sustainability is primarily cultural, not scientific or technological.

In a review of climate change communications research, Susanne Moser (2016) highlights specific psychological defences identified by researchers that can be triggered through different frames. Most relevant here is the defence that arises against identity change, a resistance to changing how we see ourselves through "avoidance, denial, helplessness, reinforcement of existing identity, or attack on others" (p. 355). The triggering frames are those that have already been used by business in shaping a corporate culture, including proclamations that

> 'The American way of life is not up for debate' or 'The threat to mobilize around is what "they" propose as solutions to climate change; Fostering anti-science and anti-government sentiments; Emphasis on freedom from government, individual freedom, [and] free market economics.'
>
> *(p. 355)*

Appeals to freedom, then and now

These threats are not hypothetical, as business has repeatedly exploited issue frames appealing to freedom. Grace Nosek (2018) highlights how the tobacco industry used this framing decades ago to protect against regulation. An industry messaging memo recommended the mantra "Freedom of choice is an American birthright. Infringement on this right is an injustice" (pp. 758–759). This argument successfully defended these businesses in the first waves of legal challenge, until the anti-tobacco movement finally reframed smoking as a systemic public health issue.

We can see explicit freedom frames already used by business in climate change litigation. When a coalition of state attorneys general recently sued fossil fuel companies for documents about whether they lied to the public and shareholders about the risks of climate change, Exxon countersued and used a freedom frame, asserting that "The allegations repeated today are an attempt to limit free speech" (Nosek, 2018, p. 767). Noteworthy also was the amplification of the corporate message through the vast communications system discussed earlier. Exxon was painted as the freedom-defending victim through two opinion pieces in the Washington Post and dozens of stories from the Wall Street Journal, Fox News, and the Heritage Foundation. The counterattack also included denunciations from other state attorneys general and threats of a counter-investigation by corporate politicians.

This is war

Because the SSE has not yet become an explicit goal of mainstream climate activism, we have not yet seen the most vicious expressions of corporate self-defence. Ron Arnold, long-time vice president of the Center for the Defense of Free Enterprise (CDFE), makes the nature of his work very clear.

> Our goal is to destroy, to eradicate the environmental movement … We're dead serious—we're going to destroy them … People in industry, I'm going to do my best for you. Environmentalists, I'm coming to get you … We're out to kill the fuckers … We [CDFE] created a sector of public opinion that didn't used to exist. No one was aware that environmentalism was a problem until we came along.
>
> *(as cited in Higgs, 2016, p. 234)*

Some who defend corporate power openly treat any attempts to regulate industry as war. Activists must prepare for the unprecedented cultural onslaught that will be waged when corporations are faced with an existential threat.

Summary

Groomed to be consumers, US citizens place disproportionate emphasis on the things they buy to attempt to shape their identity. Left unquestioned, this fact can be turned against the movement towards a SSE. But, Clive Hamilton (2010) reminds us, "identities that can be forged from the products provided by the market are not to any great degree the creations of those who adopt them, but are manufactured by marketers or popular culture" (p. 573). Aside from constraining authentic human development, this process causes a host of social ills.

> The inability of consumerism to allow true realisation of human potential manifests itself, to an ever-increasing degree, in restless dissatisfaction, chronic stress and private despair, feelings that give rise to a rash of psychological disorders—including anxiety, depression and substance abuse—and a range of compensatory behaviours including many forms of self-medication.
>
> *(Hamilton, 2010, p. 572)*

Activists must illuminate these outcomes of a society defined by consumerism.

As freedom is invoked in discussions of social issues, the effect is to legitimize or delegitimize a particular point of view or policy. As activists expose the narrowness of *unlimited consumption as freedom*, and the way it obscures the public's inability to participate in shaping the economy itself, the credibility of both the current hierarchical economic system and its defenders is undermined. It becomes harder to attack the movement to establish a SSE as this shift takes place. When citizens hear the defenders of corporate power speak about freedom and still take them seriously rather than thinking immediately of the domination they actually represent, activists have not yet done enough to clarify reality.

Redefine freedom to drive the movement

The eco-democratic definition of freedom is not only part of the cultural foundation for a SSE, which is crucial to humanity's survival, but also points towards a more humanly fulfilling society. This definition gains legitimacy as activists highlight these merits and vocally assert the concept in public discussions. By helping the public to see the climate movement as a vehicle for this kind of freedom – including, in particular, the fundamental freedom of ordinary citizens to reshape the economy to avoid ecological collapse – activists can fuel the movement.

Discourse researchers follow the ways that climate change is discussed and observe the impacts on citizen engagement with the issue. Of particular interest are depoliticizing discourses that encourage apathy. "Depoliticization refers to the deletion of alternatives and of democratic debate about alternatives regarding climate change from public spheres," notes Anabela Carvalho (2018, p. 6). "In spite of climate change's massive impacts on citizens around the world, it has been transformed into a seemingly consensual techno-managerial matter where citizens have no say. Those depoliticization processes have crucial implications for public engagement" (p. 6). How we talk about climate change determines how we understand it, and whether and how we take action. These discourses tend to be unexamined and exert an unseen marginalizing influence.

Carvalho, van Wessel, and Maeseele (2017) describe several types of marginalizing discourses. Scientization suggests that climate change is a problem of technology whose solution must be led by technical experts. Economization envisions climate change as a problem of economic calculation that can be solved entirely through market-based mechanisms, with economists at the helm. Moralization frames the issue as humanity versus CO2 and treats the solution as a matter of individual responsibility, as if personal consumption choices can solve systemic problems.

By vocally championing the eco-democratic definition of freedom and establishing the climate movement as a fight for collective self-determination within ecological limits, activists can politicize climate change and maximize mobilization through various mechanisms: issue tangibility, engaging values, movement legitimacy, and participant identity-formation.

Issue tangibility

Dale Jamieson (2017) writes about the connection between the British anti-slavery movement and current attempts to address climate change. He highlights how the distance between the colonies where slavery was practised and Britain's mainland made the issue just as distant and abstract in the minds of the public. Abolitionists realized that in order to generate a movement against slavery, they would have to make the issue visible. After parliament rejected abolition, activists started the "blood sugar" campaign – using pamphlets, speeches, and formal organizations to inform citizens about the amount of flesh they were consuming as they ate slave-produced sugar. As a result, 300,000 people boycotted Britain's largest import; abolition followed.

Jamieson (2017) observes that "For people to support moral change in a world in which there is a rupture in space, time, or scale between a cause and a harm, they must somehow be reconnected in people's consciousness" (p. 181). However, "carbon's assault on what it is to be a person seems less deep, direct, visceral and even true than slavery's assault on our shared notions of humanity" (p. 181). Highlighting climate change as a freedom issue may make it more tangible, exposing the link between this global problem and citizens' everyday experience of lacking control over their economic conditions. As the public associates the climate crisis with inadequate wages, crushing debt, and overwork – economic oppression stemming from an economy that serves elites rather than the people – the connection may be visceral enough to drive action. Achieving economic democracy would not only allow citizens to create a SSE but also address these social ills, a connection activists must emphasize.

Engaging values

Literature on climate change engagement often focuses on the connection to human values, and advocating for eco-democratic freedom may attract new movement participants for whom freedom is a highly cherished value. Corner, Markowitz, and Pidgeon (2014) note that divides on climate change are more a representation of the values at stake than disputes about science, and that if the self-transcending values typically associated with activist engagement can be joined with more traditionally self-interested values, the combination could hold promise for generating action:

> The challenge for climate change communicators seeking to make the most effective use of research on human values is to identify ways of bridging between the diverse values that any given group of individuals holds and the values that are congruent with a more sustainable society [including] coupling, for example, values around security or freedom with self-transcending values like concern for the welfare of others.
>
> *(p. 417)*

A sustainable economy will only be brought about through public control. By highlighting the freedom dimension of climate activism, activists add the promise of liberty to the range of values that inspire people to join the movement. In terms of audience, the point here is not to try to convince sceptics and "conservatives" that they should support climate action, though that could result, but to attract those already inclined to act by vocally championing a new vision of freedom.

Movement legitimacy

As the climate movement's goal of advancing a new vision of freedom becomes well known, it stands to gain broader cultural legitimacy that can also encourage participation. Julie Ayling (2017) writes that "legitimacy is an intangible but crucial resource that assists an organization to exercise authority in numerous ways" (p. 351), and the battle for legitimacy "plays a significant role in progress on climate change mitigation" (p. 366).

This is evident in the effects of the fossil fuel divestment movement, which aims to delegitimize investments in coal, oil, and gas by pointing out the need to halt carbon emissions and the industry's political efforts to block that outcome. Ultimately, activists' goal is to remove the fossil fuel industry's social license to exist. "For the divestment movement," Ayling (2017) observes, "legitimacy strengthens its case for fundamental economic change and enables it to mobilize its supporters, garner public support, and fund its activities" (p. 355). Since it began in 2012, the movement has spread to over one thousand institutions and divestment activists have transformed discussions around climate change: asserting it as a present emergency, spotlighting the culpability of the fossil fuel industry, and undermining the legitimacy of both the industry and investing in it. Ayling finds that "there are signs that the movement is making progress in building its own legitimacy and in damaging the industry's"

(p. 350). Noam Bergman (2018) notes how divestment activists have generated politicized understandings of the climate issue, which have led to increased engagement: "The most prominent impact has been the discourse shift, a clear cultural impact, which in turn has precipitated mobilisation, political and financial impacts" (p. 12).

The same can be true for climate activists aiming to redefine freedom. By asserting the legitimacy of eco-democratic freedom and establishing the movement as a vehicle for this vision of freedom, activists can legitimize their movement – encouraging participation – while also exposing the illegitimacy of an economy that is run by elites, driving towards ecological collapse, premised on limitless consumption, and forced to manufacture consent for consumerist lifestyles. As the movement gains greater social acceptance, the effect will be greater mobilization.

Participant identity-formation

Carvalho et al. (2017) cite an extensive US survey (Roser-Renouf, Maibach, Leiserowitz, and Zhao, 2014) which

> showed that 'identity' was the largest barrier to engagement with climate change politics, with a third of respondents saying they were not 'activists.' The survey also showed that most people have low expectations for the efficacy of their political actions: they do not believe that they can alter the course of climate policies and hence do not even try.
>
> *(p. 127)*

Here we see a clear indication of the connection between identity and participation in political action, suggesting that cultivating an activist identity among the public could boost engagement. Also present is a sense of fatalism. By framing the climate movement as a fight for freedom – requiring no special qualifications to participate – and informing the public that freedoms have historically been won by social movements composed of ordinary people, such self-defeating beliefs can be combatted. And by helping prospective participants see themselves as freedom fighters, activists can encourage the development of activist identities that drive engagement and commitment. These strategies also align with observations by Moser (2016) that identity-based resistance to change can be overcome through "inspiration," "appeal to deeply held values," the illustration of "new social/cultural norms," and "stories of positive transformation" (pp. 35–36).

As activists more fully develop their identity through their efforts against climate change, the movement becomes more resilient and autonomous and more compelling to those searching for an authentic self. Climate activists must be aware of the fact that, at its best, their work creates a new identity within people, and they ought to be as conscious and encouraging of that process as possible.

Summary

This section can be summed up by an activist from Toronto's climate movement, who observed that "success will be based on how much we can inspire people to engage, and how much it will feel like ... feel for people as though we were creating a space to be free" (Del Rio, 2017, p. 59). The fight for freedom has often been a central motivation of social movements. The energy behind the cause of self-determination has often led individuals to risk their lives. It lends itself to something larger than oneself, something so valuable that it can lead to the most selfless commitment, even as it allows for perhaps the most authentic expression of the self. Within mainstream climate change activism, discussions about the meaning of freedom and about the movement as a vehicle for expanding liberty have been uncommon thus far. Activists have yet to harness the force that is unlocked when establishing climate change as an issue of freedom.

Conclusion: Illuminating our choices

How can activists stimulate a society-wide reconsideration of freedom? First, activists' analysis must incorporate the necessity of a SSE in addressing climate change, which reveals the need for a cultural transformation that replaces consumerist freedom with an ecological and democratic definition. Then, campaigns dedicated to this transformation should be launched. Developing activist-controlled education and communication networks to be in constant touch with the public also seems essential to drive this cultural change and counteract the corporate culture machine. Academics writing about ecological freedom could make their work accessible to movements by hosting public discussions. All of this should be a part of larger efforts to create a participatory, democratic culture in which being intellectually and politically active is the norm, thus laying the groundwork for public control of the economy.

The nature of choice is often limiting (consider Hensher, 2019, in this collection). Defining the course of one's life, for example, means foreclosing multiple possible paths in order to follow the ones we choose. Right now we are exchanging many paths and cherished things, many freedoms, in order to pursue limitless consumption. But this is no conscious choice. In fact, it is a choice made for us by those in power, who have put immense effort into preventing us from seeing what options we have. Only when we become conscious of the trade-offs we are making will freedom of choice be anything more than an illusion. The task of activists is to make our choices clear – to help the public see that it need not walk the path to ruin. When the people recognize that they have alternatives, that the choices available are ultimately constrained only by our ongoing acceptance of elites' hierarchical and consumerist worldview and the boundaries they've established, then a transformation may be possible. We must choose to create an economy that does not force us to remain on the consumption treadmill, one that allows for the rapid emissions reductions needed to maintain a liveable climate. It is this conscious choice that will see humanity define itself, a choice that many did not know they had, now perhaps the most important one we'll ever make.

References

Anderson, K. (2015). "Duality in climate science." *Nature Geoscience, 8*: 898–900. doi:10.1038/ngeo2559.

Ayling, J. (2017). "A contest for legitimacy: The divestment movement and the fossil fuel industry." *Law & Policy, 39*(4): 349–371. doi:10.1111/lapo.12087.

Bergman, N. (2018). "Impacts of the fossil fuel divestment movement: Effects on finance, policy and public discourse." *Sustainability, 10*(7): 2529. doi:10.3390/su10072529.

Bernays, E. L., & Miller, M. C. (2005). *Propaganda*. Brooklyn, NY: Ig Publishing.

Brown, P. G. (2012). *Ethics for Economics in the Anthropocene*. Retrieved from https://therightsofnature.org/wp-content/uploads/pdfs/Peter%20Brown%20Ethics%20for%20Economics%20in%20the%20Anthropocene.pdf

Brulle, R. (2014). "Institutionalizing delay: Foundation funding and the creation of U.S. Climate change counter-movement organizations." *Climatic Change, 122*. doi:10.1007/s10584-013-1018-7.

Carvalho, A. (2018). *Discourses for Transformation? Climate Change, Power and Pathways to the Future*. Retrieved from http://repositorium.sdum.uminho.pt/handle/1822/55377

Carvalho, A., van Wessel, M., & Maeseele, P. (2017). "Communication practices and political engagement with climate change: A research agenda." *Environmental Communication, 11*(1): 122–135. doi:10.1080/17524032.2016.1241815.

Corner, A., Markowitz, E., & Pidgeon, N. (2014). "Public engagement with climate change: The role of human values." *Wiley Interdisciplinary Reviews: Climate Change, 5*(3): 411–422. doi:10.1002/wcc.269.

Del Rio, F. (2017). *In a World Where Climate Change Is Everything...; Conceptualizing Climate Activism and Exploring the People's Climate Movement* (Thesis). Retrieved from https://macsphere.mcmaster.ca/handle/11375/22508

Dibley, B. (2012). "'The shape of things to come': Seven theses on the anthropocene and attachment." *Australian Humanities Review, 52*. doi:10.22459/AHR.52.2012.10.

Dunlap, R. E., & McCright, A. M. (2011). *Organized Climate Change Denial*. doi:10.1093/oxfordhb/9780199566600.003.0010.

Dunlap, R. E., & McCright, A. M. (2015). *Challenging Climate Change: The Denial Countermovement*. Retrieved from https://www.oxfordscholarship.com/view/10.1093/acprof:oso/9780199356102.001.0001/acprof-9780199356102-chapter-10

Ewen, S. (1976). *Captains of Consciousness: Advertising and the Social Roots of the Consumer Culture* (First McGraw-Hill Paperback Edition). New York: McGraw-Hill.

Ewen, S. (2003). *PR! A Social History of Spin* (1. ed., [Nachdr.]). New York, NY: Basic Books.

Fuss, S., Canadell, J. G., Peters, G. P., Tavoni, M., Andrew, R. M., Ciais, P., ... Yamagata, Y. (2014). "Betting on negative emissions." *Nature Climate Change, 4*: 850–853. doi:10.1038/nclimate2392.

Green, F. (2018). "Anti-fossil fuel norms." *Climatic Change, 150*(1): 103–116. doi:10.1007/s10584-017-2134-6.

Hamilton, C. (2010). "Consumerism, self-creation and prospects for a new ecological consciousness." *Journal of Cleaner Production, 18*(6): 571–575. doi:10.1016/j.jclepro.2009.09.013.

Hensher, M. (2019). "A beginners guide to avoiding bad policy mistakes in the anthropocene." In *Liberty and the Ecological Crisis.* Oxford: Routledge.

Higgs, K. (2016). *Collision Course: Endless Growth on a Finite Planet.* Cambridge, MA: MIT Press.

Humboldt, W. von. (1993). *The Limits of State Action* (J. W. Burrow, Ed.). Indianapolis: Liberty Fund.

Jamieson, D. (2017). "Slavery, carbon, and moral progress." *Ethical Theory and Moral Practice, 20*(1): 169–183. doi:10.1007/s10677-016-9746-1.

Jennings, B. (2015). "Ecological political economy and liberty." In *Ecological Economics for the Anthropocene: An Emerging Paradigm* Peter G. Brown and Peter Timmerman, Eds.; (pp. 272–317). New York: Columbia University Press.

Lambacher, J. (2009). *Limits of Freedom and the Freedom of Limits: Responding to the Extinction Crisis with Responsibility, Restraint, and Joy* (SSRN Scholarly Paper No. ID 1451845). Retrieved from Social Science Research Network website: https://papers.ssrn.com/abstract=1451845

Mill, J. S. (1909). *Principles of Political Economy with some of their Applications to Social Philosophy* (7th ed.; W. J. Ashley, Ed.). London: Longmans, Green and Co.

Moser, S. C. (2016). "Reflections on climate change communication research and practice in the second decade of the 21st century: What more is there to say?" *Wiley Interdisciplinary Reviews: Climate Change, 7*(3): 345–369. doi:10.1002/wcc.403.

Nosek, G. (2018). "Climate change litigation and narrative: How to use litigation to tell compelling climate stories." *William & Mary Environmental Law and Policy Review, 42*(3): 733.

Roser-Renouf, C., Maibach, E. W., Leiserowitz, A., & Zhao, X. (2014). "The genesis of climate change activism: From key beliefs to political action." *Climatic Change, 125*(2): 163–178. doi:10.1007/s10584-014-1173-5.

Solomon, S., Greenberg, J. L., & Pyszczynski, T. A. (2004). "Lethal consumption: Death-denying materialism. In *Psychology and consumer culture: The struggle for a good life in a materialistic world* Tim Kasser and Allen D. Kanner, Eds.; (pp. 127–146). doi:10.1037/10658-008.

Stern, N. (2007). *The Economics of Climate Change: The Stern Review.* Cambridge University Press.

From navigating the Anthropocene to being in the Ecozoic

PART IV

From navigating the
Anthropocene to being
in the Ecozoic

14

A BEGINNER'S GUIDE TO AVOIDING BAD POLICY MISTAKES IN THE ANTHROPOCENE

Martin Hensher

Introduction

This chapter seeks to examine how public policy (that is, the consciously articulated statement and enactment of collective choices) might best safeguard agency and liberty in the Anthropocene era. This will be achieved by applying the lens of *optionality* to a range of different ideological responses to the challenge of the Anthropocene. This analysis does not seek to assess or appraise which of these competing visions for the future is "best" in technical terms, or even which might be most desirable in moral and ethical terms. Rather, differing proposals for safeguarding humanity's future in the Anthropocene will be contrasted, and these contrasts used to illustrate more general principles and dilemmas, with particular regard to human agency and liberty.

Given this volume's overall focus on agency and liberty in the Anthropocene, limited time will be spent on defining these terms. However, this chapter focuses on public policy and collective choice, rather than purely upon individual notions of choice and freedom. This is uncontroversial in respect of agency – "possessing the capacity to choose between options and ... being able to do what one chooses" is intuitively as applicable to single individuals as it is to groups of humans or to humanity as a whole (Horn, 2005, p. 18).

It is rather harder to avoid controversy with respect to liberty. However, in examining alternative ideological responses to an inherently collective problem, this paper follows Kitcher's "Mill amended by Dewey" communal conception of human freedom, wherein he states: "People committed to joint action with others must not impede, nor be impeded by, the similar joint projects of others" (Kitcher, 2016, p. 13). Such a communal idea of freedom reflects a partial return to conceptions of liberty that would have been more familiar in pre-modern times (Quilley, Chapter 15, This Volume). Meanwhile, optionality is intimately linked to both agency and freedom, as "the quality of being optional; opportunity or freedom of choice" (OED, 1993, p. 2011) or "the quality of being available to be chosen but not obligatory" (OED, 2019).

The origins, evolution, and characteristics of the Anthropocene era are described elsewhere in this volume. What is relevant to the analysis presented in this chapter are alternative human responses to the reality of the Anthropocene – the different trajectories and routes that have been proposed to carry humanity away from the gravest dangers implicit in an Anthropocene world, namely, profound human suffering, the stunting of human flourishing, and the potential for ecological and civilizational collapse – the "common ruin of all" (Angus, 2016, p. 222). The arrival of the Anthropocene itself represents a paradox of human agency – humanity has been able to change the very functioning of the earth system to such an extent that it may threaten the continuity of modern civilization. Clive Hamilton (2017) suggests that the Anthropocene represents a time of narrowing limits to agency, which will confront humanity with our inability to control our common fate. The analysis that follows will contrast a range of strategies that seek actively to control our destiny through enhanced technology and management, with others that seek alternative paths of "meekness" (Hamilton, 2017, p. 9), and then attempts to draw out what they might mean for the concepts of agency and liberty described above.

Optionality and choice in the Anthropocene

It is the very essence of the Anthropocene that it constrains certain human choices through irreversibility. The Anthropocene has already arrived. It is a state, not a choice. A choice can no longer be made to *return* to Holocene conditions to avoid a warmed Earth and a changed biosphere. Steffen et al. (2018) argue that choices now only exist about future pathways, effectively between choosing a trajectory that keeps warming below 2°C under a "Stabilized Earth" pathway, or slipping across the threshold onto a "Hothouse Earth" pathway, with all the dire consequences that would entail. Deep uncertainty exists regarding the thresholds and feedback loops that might differentiate these trajectories; on their consequences for humanity; and on the effectiveness and feasibility of different courses of action in steering human action towards one or other of these trajectories. Nevertheless, what is at stake is clear – avoiding ecological collapse, the consequent collapse of modern human civilization, and the suffering and loss of life (both human and non-human) that would potentially accompany these twin collapses. The destruction of opportunity that would accompany ecological and civilizational collapse would represent the curtailing of liberty and agency on a grand scale; constellations of possible choices that are taken for granted today would be foreclosed, while many human lives would be deprived of all liberties through being cut short or never conceived.

It is, of course, possible to argue that *civilizational* collapse (by which we might mean the passing of a dominant societal organising paradigm, be that capitalism, liberalism, "Western civilization," patriarchy, modernity, or some other appellation) might not be such a disaster, and might open spaces for new civilizations to grow which deepen human agency and freedom. Yet the *combination* of ecological collapse and consequent civilizational collapse raises the stakes substantially; its double impact raises the prospect of terrible physical and material suffering for billions, while ecological collapse might prove to be an insurmountable constraint on the scope for rebuilding new civilizations in the future. The avoidance of ecological collapse is then clearly a necessary precondition for the preservation of meaningful agency at both individual and societal levels. The maintenance of as much as possible of what is good and valuable for human flourishing within modern civilization then also becomes an important target for preservation on instrumental grounds. This is perhaps one of the "wicked dilemmas" that Kish and Quilley (2017) suggest are typical of humanity's passage into the Anthropocene: we must seek to retain some of the most important fruits of modernity (e.g. public health, essential health care, universal basic education, etc.) if we are to avoid unprecedented humanitarian disaster, yet this modernity and its liberal society "owes its existence to the bubble of ecological affluence fuelled by the 'discovery' of the New World and the exploitation of the stored solar energy in fossil fuels," while actively driving the process of ecological collapse (Ophuls, 2011, p. 130). Thus the job of public policy today – in the face of clear, present, and grave danger – is to maximize the long-term prospects of survival for the greatest possible number, while attempting to best ensure that they and their descendants enjoy meaningful choices for their own future flourishing – while recognising that the old tools of modernity and the growth economy are no longer our friends as ecological scarcity returns (Ophuls, 2011).

Conventional approaches to policy analysis and evaluation (e.g. cost–benefit analysis, decision analysis) profoundly struggle to incorporate the level of uncertainty inherent in contemplating humanity's future in the Anthropocene. They require quantified estimates of risk, cannot deal with deeper uncertainty, and struggle to capture the human cost of more extreme (yet possible) outcomes. By contrast, Taleb's concept of "optionality" (Taleb, 2012, pp. 180–186) rejects the teleological fallacy of believing we know where we are going – either as individuals or as a common humanity. Taleb argues that the wise way forward is to maximize optionality, that is, to choose options for which potential losses are limited and known, while accepting (and actively seeking) gains that are unbounded and unknown. In the Anthropocene, the path of maximising optionality would seek to minimize downside losses by assiduously avoiding ecological and civilizational collapse, and – by the very act of maintaining a viable human civilization – enjoy the unknowable benefits that will follow. The lens of optionality will be applied to the key choices facing humanity in the Anthropocene, and to a variety of the ideological programs that have been proposed as responses to humanity's Anthropocene predicament.

Competing pathways to the future

A lively debate on alternative social and economic models that might guide humanity through an era of growing environmental crisis dates back to the 1970s (Meadows, Meadows, Randers, and Behrens, 1972). Perhaps the first well-formed model or pathway to emerge was Herman Daly's "steady state economy" (Daly, 1977), under which the guiding purpose of economic policy becomes the maximization of human well-being subject to a material throughput constraint that can be sustained indefinitely. Much of the subsequent debate has centred on whether economic growth can be "decoupled" from material throughput, allowing continued "green growth" and hence "green" capitalism (e.g. Bowen and Hepburn, 2014), and whether the *absolute decoupling* that this requires is actually feasible (e.g. Ward et al., 2016). If it is not, then some form of "voluntary degrowth" will be required (e.g. Kallis, 2011); a deliberate reduction in material throughput and economic output to return humanity to a trajectory of material throughput that can be sustained without environmental collapse.

In recent years, a newer school of thought has pushed beyond the "green growth" concept. Ecomodernists (Asafe-Adjayu et al., 2015) have suggested the pursuit not only of absolute decoupling through the aggressive promotion of technological solutions (ranging from large-scale expansion of of nuclear power to the increased use of genetically modified organisms), but also the active management of the biosphere through geoengineering to mitigate the worst effects of climate change. Meanwhile, proponents of "ecosocialism" (Angus, 2016; Phillips, 2015) appear at various points along the spectrum between degrowth and ecomodernism. Self-described ecosocialists take different technical positions on what we might call "management approaches", but are united in their confidence that only the supplanting of capitalism by democratic socialism can allow the necessary steps towards sustainability to be taken (Löwy, 2007; Phillips, 2015).

All the while, the spectre of failure floats in the wings of this debate. Bonaiuti (2017) reprises the logic of the original *Limits to Growth* analysis, namely that without a significant change of direction, the default path of increasing growth and material throughput will lead to what he describes as "involuntary degrowth." This path is characterized by diminishing marginal returns to complexity across society, and the increasing inability of social, economic, and ecological systems to maintain their current levels of complexity, with grave losses of human welfare being the inevitable consequence. Others are more willing to name the likely outcome of this default path squarely as "civilizational collapse" (Angus, 2016, p. 222). Whilst most discussion of this family of outcomes not surprisingly views them as morally disastrous calamities, some at times appear to suggest that collapse could bring a form of redemptive purging and healing of humanity's rift with nature (Jensen, 2016; Olson, 2012) – albeit at the cost of billions of human lives. We might label such views as "primitivist" – they envisage a future global population of a few million humans returned to a pre-agricultural (if not pre-lapsarian) existence in a world beginning to heal itself.

While varying wildly in their prescriptions and core assumptions, all the future trajectories described above have in common the starting point that "business as usual" is not ecologically sustainable, and all seek in very different ways to address this fundamental problem. Indeed, Figure 14.1 does not attempt to locate "business as usual" in this particular space, as elements of several of the strategies described above are present to some small extent in current policies and economic models. However, readers will indulge the inclusion of the category of "denialist capitalism" in Figure 14.1. There are still those who deny the reality of humanity's Anthropocene predicament, and who seek only growth at all costs (even if only for their own benefit). Such a path needs little further analysis.

There are, however, other visions of pathways to the future which are not rooted first and foremost in the ecological crisis. Two that are of particular significance are "transhumanism," and what we might call the "cosmic expansion imperative." Transhumanism is used here as a shorthand for what O'Connell (2017) describes as the movement for human emancipation from the limitations of biology, by means of technologies such as life extension, cyborg augmentation, and brain/computer interfaces. Alternatively, space colonization is often mentioned as the only strategy by which humanity can escape terrestrial limits, and thus evade other existential risks to humanity's long-term survival (e.g. Bostrom, 2003).

Figure 14.1 attempts to plot these different pathways and futures against two dimensions: the extent to which they require increasing or decreasing economic growth and material throughput, and differing dominant forms of relationships between humanity and the natural environment. These relationships include humanity being overwhelmed by ecological breakdown; adapting to and accommodating environmental changes; actively seeking to control the natural

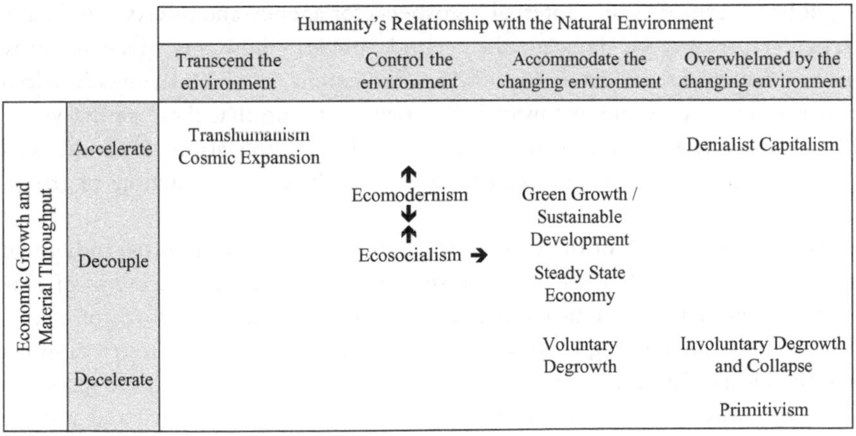

		Humanity's Relationship with the Natural Environment			
		Transcend the environment	Control the environment	Accommodate the changing environment	Overwhelmed by the changing environment
Economic Growth and Material Throughput	Accelerate	Transhumanism Cosmic Expansion	↑ Ecomodernism ↓	Green Growth / Sustainable Development	Denialist Capitalism
	Decouple		↑ Ecosocialism ➜	Steady State Economy	
	Decelerate			Voluntary Degrowth	Involuntary Degrowth and Collapse Primitivism

FIGURE 14.1   Pathways to the future – Their relationship to material throughput and the natural environment

environment (e.g. through geoengineering); or seeking to escape and evade environmental limits. For example, the pathway of space colonization is likely to require an "accelerationist" doubling down on growth and technology (to get a non-trivial number of humans off the Earth in the first place), in order to directly escape the ecological and resource constraints of Earth's planetary boundaries. Similarly, degrowth requires a "decelerationist" reduction in material through-put to ecologically sustainable levels, while accepting that human society will need to adapt to and accommodate irreversible environmental changes already set in motion by past human activity. Particularly interesting here is the tension within ecomodernism between its clearly articulated desire to drive ecological decoupling through technology, and its parallel desire to control the environment through active planetary management interventions such as geoengineering; this suggests a great deal of fluidity in what "ecomodernism" might actually entail, with the potential for tempting (but risky) trade-offs between decoupling and environmental control.

Agency, freedom, and optionality

Taleb's conception of optionality is relevant to the present volume's analysis of agency and liberty in the Anthropocene in a very simple and direct way. It advocates for the preservation of freedom of action in an environment of deep uncertainty, through careful choices that avoid catastrophic errors and downside risks. In so doing, it maximizes the space available for unknown but potentially unbounded future benefits. How, then, might the attempt to preserve and maximize optionality help us to assess the differing prospects for agency and freedom embodied in those pathways that seek to avoid and prevent collapse?

The relationship between each of these "pathways to the future" with agency, liberty, and optionality can be considered in two ways. The first concerns what might be described as the "internal" prospects for agency and liberty should any particular pathway be pursued; the second concerns how any given pathway might interact with or cut across others – the extent to which there is freedom to pursue more than one pathway. It is worth restating that these pathways are inherently collective endeavours – each will have implications for individual agency and liberty, but all are meaningful only as the collective actions of groups or societies.

Assessing how each of the pathways might support or undermine individual freedom is surprisingly difficult to do in the abstract. A few statements of tendency can be made without requiring too much imaginative effort: for example, ecosocialism and degrowth are likely to require significant restrictions on private ownership and accumulation, while green growth seeks to preserve such economic freedoms by "dematerializing" economic activity and profits. Alternatively, a species survival strategy that focused on space exploration to establish off-planet human settlements as rapidly as possible might be quite likely to involve a rather austere and single-minded mission focus, which had limited time for several currently important aspects of personal freedom, whether off-world or back on Earth. Yet without being able to map out a whole institutional and political architecture for them, it is all but impossible to compare, for example, the implications for individual freedom of ecomodernism versus a steady state economy. Almost any of the pathways could be pursued under very different political and institutional arrangements, and under very different cultural imperatives and conceptions of the good life. Most of the pathways could possibly be made conducive to individual liberty within wider societal and ecological constraints; by the same token, history suggests that any of the pathways could reliably be turned into thoroughly unpleasant and tyrannical dystopias if the right personnel were invited to apply for senior management roles. As a result, it is currently more useful to understand how the pursuit of different pathways might impede that of others.

The first class of interactions or conflicts between the competing pathways that requires consideration involves *ruin problems* (Taleb, Read, Douady, Norman, and Bar-Yam, 2014) – generalized, irreversible, and catastrophic harms, which cannot be recovered from and hence have effectively infinite costs. Clearly, most of the pathways summarized above might be seen as seeking, in some measure, to lead humanity safely *away* from one class of ruin problem, namely ecological and/or civilizational collapse. Taleb et al. suggest that ruin problems are precisely the category of problem for which the *precautionary principle* can appropriately be invoked. Even with only very small probabilities, the effectively infinite costs of ruin mean that the risk of incurring ruin must be avoided at all costs; no potential upside benefit can justify incurring this risk. Two of the potential pathways to the future have elsewhere been identified as carrying with them potential ruin risks: aspects of the transhumanist agenda, and parts of the Ecomodernist program. The potential dangers of the emergence of uncontrollable Artificial Intelligence (AI) and "superintelligence" have been categorized by Bostrom (2014) as having the potential to pose an existential risk to humanity. While the transhumanist program (in all its diversity) does not necessarily entail the development of superintelligent AI, its pursuit certainly makes such an eventuality more, rather than less, likely than under other developmental pathways. Meanwhile, Ecomodernists have called for the deployment of two technologies which others have decried as risking potential ruin: the widespread use of genetically modified organisms (GMOs) and the use of geoengineering. GMOs are the very example that Taleb et al. use as a ruin problem in which they argue the precautionary principle should indeed be applied to prevent their use; while geoengineering has been argued to entail similar risks of extreme unintended consequences (Lent, 2017).

One of the key reasons ruin risks are to be avoided is that they are effectively *irreversible*. But even in the absence of ruinous outcomes, irreversibility is potentially a problem for the maintenance of meaningful choice and agency. The risk of ecological overshoot and system collapse is a problem, in part, because it is effectively irreversible in the short to medium term. Recovery of technological and institutional functionality after a widespread collapse may not be possible, due to resource depletion incurred during the modern era; the depletion of easily extractable fossil fuel resources is well advanced, and ever harder-to-reach reserves may become effectively unavailable to a future society with reduced technological and energetic resources at its disposal. Indeed, "recovery" from a thorough-going collapse resulting in a primitivist scenario may not be socially or institutionally possible on timelines less than centuries or even millennia. Conversely, accelerationist pathways may entail significant risks of irreversible outcomes. Not only might the emergence of superintelligent AI be an irreversible outcome with enormously unpredictable consequences; less extreme "progress" in human augmentation and life extension might create a new class of transhumans with extremely high material, medical, and technological needs, which could not ethically be denied once instantiated. One of the risks of geoengineering is that some techniques might similarly display irreversible features. Irreversible outcomes do not need to have catastrophic consequences for them to impinge upon human agency; their very existence restricts choice by preventing the possibility of retreat or by reducing room for manoeuvre.

Perhaps a broader challenge to collective agency and societal freedom to choose is a strong sense that a number of the alternative pathways might ultimately be mutually exclusive. This requires consideration of whether alternative pathways could be pursued simultaneously without coming into irresolvable conflict. This problem may have different manifestations at individual, local, national, or global levels, and may stem either from collective action and coordination problems, or from the tendency of some pathways to dominate others. For example, individuals or communities might be able to pursue aspects of a degrowth strategy in their social and economic choices, even while coexisting within a society which is following a different path; some contemporary "degrowthers" advocate and follow this approach. At larger scales, however, such an approach of parallel or alternative ways of living may fail to deliver on one of the key purposes of degrowth – namely the avoidance of ecological collapse. Unless a sufficient (and large) proportion of the population were to adopt degrowth habits, it is unlikely that their efforts would scale adequately to impact upon the overall global growth/degrowth trajectory. Indeed, in the extreme, it can be argued that only a fully globally coordinated degrowth strategy could guarantee success, as there are strong collective action problems and incentives conspiring against spontaneous and selfless action on the part of individuals, households, firms, and nations. This is, of course, a familiar argument from attempts in recent decades to combat anthropogenic climate change, and is often used by climate sceptics as an argument against action; sadly, this does not mean there is not some truth in it. The voluntary degrowth, steady state, and green growth pathways could all be thwarted if significant parties (whether nations or corporations) refused to comply and continued to pursue denialist capitalism. A strong national or global consensus to pursue any sustainable pathway with real vigour (and hence any real chance of success) must entail strong enforcement of policies and regulations, and consequent restrictions on the freedom of individual or corporate actors to undermine the selected policy path. There is no shying away from the fact that avoiding involuntary degrowth and collapse necessarily entails restrictions on economic freedom, and is probably not compatible with current forms of capitalism.

By contrast, certain other pathways might display a strong tendency to *dominate* others once embarked upon at any scale, especially amongst pathways falling in the "accelerationist" space. Certain elements of the Ecomodernist agenda might have very substantial externalities which would affect all people everywhere, whether or not they had signed up to the project – most obviously in the form of geoengineering, which would essentially be global in scope. This may or may not prevent individuals or communities from pursuing other local pathways: for example, a region or even a nation could choose to pursue elements of a voluntary degrowth strategy while others pursued an Ecomodernist path, but no nation could choose to opt out of living in a geoengineered world, if that is the path others have taken. Such a nation or community might be tolerated as having made an idiosyncratic choice in favour of abstinence; yet it might come to be viewed as an irritant, a free-rider, or even a threat if it sought to prevent the Ecomodernist policies of other nations from impacting on it. It is also not hard to imagine the transhumanist pathway coming to dominate all other options, in more or less dramatic ways. Human–machine integration and augmentation could have very unpredictable consequences for the individual's freedom even to be "human," long before any putative "Singularity" might be reached. The emergence of strong artificial intelligence (Bostrom, 2014) could close down options for decelerationist pathways if the AI offered up accelerationist alternatives that had not previously been available. Human attempts to achieve the goal of superintelligent AI could by themselves drive accelerating energy and resource use that undermined the decelerationist plans of others (Bratton, 2015) – yet it is entirely possible that ever-better AI could assist in making a reality of decoupling and avoiding the need for deceleration. Sadly, it is also likely that nations or groups pursuing accelerationist strategies of all types might be prone to exerting geopolitical dominance over communities or nations attempting to follow other pathways to sustainability. All accelerationist strategies are likely to covet the natural resource endowments of other nations. To accelerationists, the decision of a nation pursuing (say) a steady state economy to leave its resources in the ground will not just be an inconvenience, but a provocation, if not a heresy. Ironically, the Anthropocene may provide ample opportunities for highly traditional conflicts between national liberation and domination to play out long into the future.

The preceding discussion also hints at a difficult tension that may impede the freedom of nations or societies to pursue decelerationist pathways. This is the question of whether and how it might be possible to pursue a decelerationist trajectory for consumption and material throughput, while retaining access to key advanced technologies. Some critics of degrowth, including ecosocialists such as Phillips (2015) draw a caricature of a hair-shirted "year zero" philosophy that revels in austerity and rejects all "modernity" out of hand. Most steady state or degrowth advocates do not have a doctrinal objection to all technology, but degrowth in particular can be vulnerable to criticism in this vein. To achieve a welfare-enhancing deceleration, rather than a welfare-destroying regression, some aspects of a dual economy (Boeke, 1953) are unavoidable, with even a degrowth economy needing to maintain aspects of advanced manufacturing and services (e.g. medicines and medical technology, computing and artificial intelligence). Indeed, in a difficult irony, advanced defence and security capabilities may be required in order to protect the very choice of degrowth in a world in which other nations have chosen different pathways. Freedom to choose may have to be heavily hedged by pragmatism, with the inherent risk that the measures required to safeguard "degrowth in one country" might undermine the purity of this very choice in practice.

At the opposite end of the spectrum, there is increasing concern that the pursuit of the transhumanist agenda may lead not just to dualism and ever-increasing inequality, but to an effective bifurcation of humanity into two species following a "techno split" (Lent, 2017) – "natural" humans and augmented "post-humans" (Harari, 2016). Such a divergence would raise complex and novel questions of agency, freedom, and rights. Managing the conflicting choices of different human tribes has proved challenging enough throughout history; managing the choices of two increasingly divergent intelligent species without one becoming subordinate to the other could be even harder.

A final risk to optionality and maintaining real freedom to choose different pathways arises from the potential for "moral hazard." The pursuit of certain pathways may undermine the incentives to allow for or to preserve future freedom to choose other paths. This is most obviously visible in the potential pursuit of space colonization as a strategy for human survival in the Anthropocene. Excessive focus on creating an escape into space – especially if it took the form of what Harari (2016) calls an "elite ark" (i.e. by which ruling elites or their progeny would be the ones to leave the Earth) – could lead inexorably to a failure to take the steps required to safeguard the future of those left behind on Earth. Similarly, Lent (2017) argues that the chief risk of geoengineering is precisely that of moral hazard – faith in an untestable technology may be used as an excuse not to take the steps needed to reduce consumption. In the opposite direction, Deep Green devotees of a return to hunter-gatherer primitivism clearly have little incentive to protect the interests of those with differing views. While certain pathways might possibly allow for some indigenous peoples to choose a more traditional way of life (by no means the same thing as "primitivism"), this is an option that is unlikely to be available to non-indigenous peoples at scale in any scenarios that avoid collapse. For the primitivist, civilizational collapse may be something to be welcomed; which is not then too many steps from something that they might have an interest in hastening on its way.

Conclusion: Safeguarding optionality in the anthropocene

The discussion above explains why retaining and maximising optionality is essential to protecting human agency in the Anthropocene era. Optionality prizes the avoidance of known or knowable catastrophic errors and downsides or ruin risks, so that the potential for future benefits might be maximized. Viewed through the lens of optionality, it is possible to offer some tentative conclusions regarding the risks inherent in the different pathways reviewed in Figure 14.1. It is not possible to offer absolute guarantees that decelerationist pathways (degrowth or SSE) will necessarily avert ecological collapse. Yet accelerationist pathways (green growth, ecomodernism) bring with them a much greater inherent risk of ecological collapse – they depend entirely on the untested assumption that absolute, sufficient decoupling can be achieved (and/or direct control of environmental processes achieved through geoengineering). They increase downside risks substantially, and (under ecomodernism) flirt explicitly with ruin risks. They would introduce additional layers of complexity and fragility (increasing their chances of failure), while decelerationist pathways offer the opportunity to reduce the complexity of social and economic systems, and to nurture Taleb's quality of *antifragility* in systems and institutions that can thrive in volatility and disorder. Indeed, done well (an important qualification), degrowth conforms strongly to Taleb's method of the *via negativa*: "to take away from the future, reduce from it, simply, things that do not belong in the coming times" (Taleb, 2012, p. 310). Meanwhile, transhumanism and cosmic expansion carry two great risks as future pathways for humanity. Neither entails any direct consideration of the ecological crisis; relying on them for its eventual resolution therefore seems entirely unreasonable. They are also proceeding quite separately from attempts to resolve the ecological crisis, and they are being pursued increasingly by private interests; alongside their inherent risks of moral hazard, they also contain the seeds of exponentially accelerating inequality between humans. Their potential to undermine or confound future efforts to deliver global social, economic, and ecological solutions should not be underestimated.

Perhaps the greatest challenge facing us as we weigh alternative pathways through the Anthropocene is to avoid the conflation of means with ends, or – more specifically – the conflation of technical options with ideological worldviews. *Total ideologies*, which proceed from a system of socially determined and "apparently unquestionable assumptions" (Popper, 2003 [1945], p. 237) will inexorably foreclose practical options that may have actually worked. Quilley (Chapter 2, This Volume) develops the concept of seeking the "adjacent possible," bringing together combinations of concepts and options for action in permutations which are currently unthinkable or invisible to prevailing modes of thought or ideology. A focus on optionality offers one tool for identifying ideological blind spots.

I have argued that safeguarding optionality is an essential prerequisite for protecting human agency and collective freedom to choose in the Anthropocene.

We might go further to suggest that the pursuit of optionality itself represents the constant and ongoing exercise of agency and freedom. Needless to say, this is not the same thing as arguing that either will be easy to do. The mindset that values the preservation of optionality over selecting the "right" course of action is hard to achieve, and the ability to think in these terms may even be quite rare in modern culture. Safeguarding optionality requires a judicious mix of caution and risk-taking; Taleb (2012) has an instinctive disdain for the ability of bureaucracies (public and private) and politicians to honestly identify and avoid catastrophic risks, or to take the calculated risks that can capture an effectively unlimited upside without risking ruin. Yet, in the Anthropocene, these are precisely the tasks that our societies, governments, and political systems must master collectively. Promoting deeper consideration of optionality in policy development and assessment inherently requires interdisciplinary thinking, as so many future consequences fall in domains far removed from those in which the original decision may have been taken. Preserving optionality in public policy requires a sound understanding of systems thinking and complexity, as well as a significant degree of humility (Sherden, 2011). It is increasingly clear that repairing the damage caused by decades of neoliberal ascendancy to the profession of public policy and administration will require a significant project in its own right (Mitchell and Fazi, 2017); understanding optionality and how to use it to guide decisions small, large, or existential will form a crucial part of the curriculum for future public policy practitioners in the Anthropocene.

References

Angus, I. (2016). *Facing the Anthropocene: Fossil Capitalism and the Crisis of the Earth System.* New York: Monthly Review Press.

Asafe-Adjayu, J., Blomqvist, L., Brand, S., Brook, B., Defries, R., Ellis, E., … Teague, P. (2015). *An Ecomodernist Manifesto.* Retrieved from http://www.ecomodernism.org/manifesto-english/

Boeke, J. (1953). *Economics and Economic Policy of Dual Societies, as Exemplified by Indonesia.* New York: Institute of Pacific Relations.

Bonaiuti, M. (2017). "Are we entering the age of involuntary degrowth? Promethean technologies and declining returns of innovation." *Journal of Cleaner Production.* doi:10.106/j.jclepro.2017.02.196.

Bostrom, N. (2003). "Astronomical waste: The opportunity cost of delayed technological development." *Utilitas, 15*(3): 308–314.

Bostrom, N. (2014). *Superintelligence: Paths, Dangers, Strategies.* Oxford: Oxford University Press.

Bowen, A., & Hepburn, C. (2014). "Green growth: An assessment." *Oxford Review of Economic Policy, 30*(3): 407–422.

Bratton, B. (2015). *The Stack: On Software and Sovereignty.* Cambridge, MA: MIT Press.

Daly, H. (1977). *Steady-state Economics: The Economics of Biophysical Equilibrium and Moral Growth.* San Francisco: W.H. Freeman and Co.

Hamilton, C. (2017). *Defiant Earth: The Fate of Humans in the Anthropocene.* Sydney: Allen and Unwin.

Harari, Y. N. (2016). *Homo Deus: A Brief History of Tomorrow.* London: Harvill Secker.

Horn, J. (2005). "Agent." In T. Honderich (ed.), *The Oxford Companion to Philosophy* (pp. 18). Oxford: Oxford University Press.

Jensen, D. (2016). *The Myth of Human Supremacy.* New York: Seven Stories Press.

Kallis, G. (2011). "In defence of degrowth." *Ecological Economics, 70*(5): 873–880. doi:10.1016/j.ecolecon.2010.12.007.

Kish, K., & Quilley, S. (2017). "Wicked dilemmas of scale and complexity in the politics of degrowth." *Ecological Economics, 142*: 306–317. doi:10.1016/j.ecolecon.2017.08.008.

Kitcher, P. (2016). "Masking the meaningful." *Global Policy, 7*(Supp. 1), 5–15.

Lent, J. (2017). *The Patterning Instinct: A Cultural History of Humanity's Search for Meaning.* Amherst, NY: Prometheus Books.

Löwy, M. (2007). "Ecosocialism and democratic planning." *Socialist Register, 43*: 294–309.

Meadows, D. H., Meadows, D. L., Randers, J., & Behrens, W. W. (1972). *The Limits to Growth: A Report for the Club of Rome's Project on the Predicament of Mankind.* London: Potomac Associates.

Mitchell, W., & Fazi, T. (2017). *Reclaiming the State: A Progressive Vision of Sovereignty for a Post-neoliberal World.* London: Pluto Press.

O'Connell, M. (2017). *To be a Machine: Adventures among Cyborgs, Utopians, Hackers and the Futurists Solving the Modern Problem of Death.* New York: Doubleday.

OED. (1993). *The New Shorter Oxford English Dictionary on Historical Principles, Vol. 2 N-Z.* Oxford: Clarendon Press.

OED. (2019). Oxford Living English Dictionary. Retrieved from https://en.oxforddic tionaries.com/definition/optionality

Olson, M. (2012). *Unlearn, Rewild.* British Columbia: New Society Publisher.

Ophuls, W. (2011). *Plato's Revenge: Politics in the Age of Ecology.* Cambridge, MA: The MIT Press.

Phillips, L. (2015). *Austerity Ecology and the Collapse-porn Addicts: A Defence of Growth, Progress, Industry and Stuff.* Winchester: Zero Books.

Popper, K. (2003 [1945]). *The Open Society and Its Enemies* (Vol. II: Hegel and Marx). Abingdon: Routledge.

Quilley, S. (Chapter 15, This Volume). Liberty, energy, and complexity in the longue durée." In C. J. Orr, K. Kish, & B. Jennings (eds.), *Liberty and the Ecological Crisis: Freedom on a Finite Planet.* Forthcoming.

Quilley, S. (Chapter 2, This Volume). "Liberty in the near Anthropocene: State, market, and livelihood" In C. J. Orr, K. Kish, & B. Jennings (eds.), *Liberty and the Ecological Crisis: Freedom on a Finite Planet.* Forthcoming.

Sherden, W. (2011). *Best Laid Plans: The Tyranny of Unintended Consequences and How to Avoid Them.* Santa Barbara: Praeger.

Steffen, W., Rockström, J., Richardson, K., Lenton, T. M., Folke, C., Liverman, D., ... Schellnhuber, H. J. (2018). "Trajectories of the earth system in the anthropocene." *Proceedings of the National Academy of Sciences, 115*(33); 8252.

Taleb, N. (2012). *Antifragile: How to Live in a World We Don't Understand.* London: Allen Lane.

Taleb, N., Read, R., Douady, R., Norman, J., & Bar-Yam, Y. (2014). *The Precautionary Principle (with Application to the Genetic Modification of Organisms).* New York. Retrieved from https://arxiv.org/pdf/1410.5787.pdf

Ward, J. D., Sutton, P. C., Warner, A. D., Costanza, R., Mohr, S. H., & Simmons, C. T. (2016). "Is decoupling GDP growth from environmental impact possible?" *Plos One, 11*(10): 10.1371/journal.pone.0164733. doi:10.1371/journal.pone.0164733

15

LIBERTY, ENERGY, AND COMPLEXITY IN THE LONGUE DURÉE

Stephen Quilley

Introduction

Liberty in the Anthropocene implies a geological/evolutionary timescale and suggests a comparison with other ecological epochs that is provocative and problematic in equal measure. Even the briefest consideration of the idea of liberty in the antecedent epoch, the Holocene, highlights the enormous variability of the human condition over the course of human development. At the start of this period all humans lived in small, familial bands at very low population densities. Just after the mid-way point, Egyptian Pharaohs were enslaving thousands in the cause of monumental immortality projects. And at the gloaming of humanity's adolescence, Thomas Paine proclaimed the universality of individual moral agents, sacrosanct with rights, answerable and possessed of a rich interiority, a potential for autonomy, and a greatly enhanced capacity for self-reflection. The Palaeolithic hunter, the Canaanite slave, the Quaker turned firebrand: how fraught is the attempt to articulate general propositions! To attempt to delineate the arc of freedom into a future that is radically, seismically uncertain would seem, perhaps, to be a fool's errand. However, the big history of liberty can at least illuminate some basic parameters.

The historicity of liberal ideas: Liberty and the ontology of rational individualism

Much of the argument, that I will develop here, stands in direct opposition to taken for granted assumptions underlying mainstream understandings of "freedom." Over the last 500 years, we have seen a long and apparently wide-ranging debate about the nature of liberty, its relation to the state and society, and the conditions that attach to freedom, etc. However, against our geological timeframe, such debates appear incredibly narrowly defined.

Liberals accord liberty primacy as a political value – because this reflects, as John Locke put it, some original "State of Perfect Freedom [in which people are free to act] ... as they think fit ... without asking leave or depending on the Will of any other Man" (Locke 1960 [1689]: 287). This was echoed 200 years later by J.S. Mill, who argued that the burden of proof is upon those who would limit freedom through coercive means (1963, vol. 21: 262). This fundamental liberal principle has also provided the starting point for more recent philosophers such as John Rawls (Gaus 1996: 162–166). The point of these debates has always been that political authority, and especially the use of force, requires justification – the driving problematic of all social contract theory from Hobbes, Locke, Rousseau, and Kant, through to diverse contemporary interpretations. Thus, Rawls's first principle of justice is that "each person is to have an equal right to the most extensive system of basic liberty compatible with a similar system for all." (Rawls 1999: 220) From this basis, major disagreements have turned on the extent to which freedom is to be understood as the absence of coercion by other people (Berlin's [1969] "negative freedom"); or [from Rousseau and Mill] a more positive emphasis on autonomy – the freedom to be able to act in certain specified ways (Green 1895).

At the same time, there are significant disputes about the relation between freedom, private property, and the market. The latter are seen by some classical liberals, and many libertarians, not simply as practical guarantors of liberty (Hayek 1978), but defining features without which liberty cannot exist (Robbins 1961; Gaus 1996; Steiner 1994). Such agreement notwithstanding, there are of course a range of finely nuanced views about the extent to which liberalism is compatible with taxation and state security and law enforcement. Where such concerns begin to admit arguments for the redistribution of wealth and a more activist state, classical liberalism morphs into a myriad of revisionist varieties of welfare-liberalism which hinge on a pragmatic balance between market and state, equity and growth, innovation and stability – concerns which ultimately prioritize the positive liberty of economic have-nots (Freeden 1978; Keynes 2007; Beveridge 1944).

However, the fervour, fine-grain, and intractability of these disputes really obscures the narrow application of liberal concepts. The most obvious difficulty is that they take for granted a social world dominated by the agency of rational individuals. To a great extent this methodological individualism started as a normative and ideological reflection of the new social and economic world that was brought into existence by the process of capitalist modernization and the antecedent protestant reformation. Whether Adam Smith's conception of rational preference seeking individuals with a natural propensity to truck and barter, or John Locke's articulation of individual natural rights, Tom Paine's proclamation of universal human rights, Kant's insistence on individual reason as the source or moral and aesthetic progress, or Rousseau's utopian conception of individual freedom as an original state of grace – the project of Enlightenment insisted on the naturalness and universality of free individuals equipped (by God or nature) with the faculty of reason, capable of cooperating, contracting, and advancing the human condition to the benefit of all. By the middle of the nineteenth century, Victorian philosophers and armchair anthropologists were beginning to forget the utopian and constructivist origins of Enlightenment individualism. Instead, both the history and possible futures of the world were shoehorned into the a methodologically individualist framework (e.g. J.S. Mill 1963 [Vol. 8]: 879; Jeremy Bentham 1823, Chap 1, sec. 4; Herbert Spencer 1851: 1) that has dominated the social and political imaginary of modern societies ever since.

But seen against the longue durée of human development, these liberal political-philosophical debates, and the empirical phenomena to which they refer, are indivisible from the particular complex industrial societies associated with modernity. In an individualized society, with a heavy emphasis on the autonomy and agency of wilful selves, it is perhaps unsurprising that people are drawn to the philosophically idealist notion that liberal-individualist theories of justice, morality, and social-political institutions – are natural, universally valid and independent of material circumstance; and that psychology and personality bear the heavy imprint of culture. In fact, liberal ideas of freedom are highly contingent, and emerged on the back of specific processes of capitalist modernization and state formation. The assumption of individual agency and rationality has become a necessary fiction written into the architecture of all modern institutions and disciplines, from democracy, economy, and the law, to medicine, philosophy, and religion. But it also speaks to the emerging reality of a society of socially and spatially mobile individuals.

Culture and personality

Studies in anthropology, sociology, social psychology, and neuroscience have shown conclusively that the universality of human nature (i.e. the genetic and physiological potential of people qua humanity) is emphatically not inconsistent with the variability of forms of substantive rationality and average personality between societies and through time. Pioneering studies under the umbrella of the "culture and personality" school in anthropology (Hofstede and McCrae 2004; Levine 2001; Lindholm 2001; Mead 1935; Sapir 1949; Kluckhohn and Murray 1953; Benedict 1934), more recent studies in anthropological linguistics (Frank et al. 2008), philology (Barfield 1965), and historical sociological studies by Elias (2012) have been echoed by recent research in the field of brain science (Henrich et al. 2010), all underlining a central proposition – that variable patterns of socialization associated with a wide variety of environmental and sociocultural contexts, can and do give rise to very distinctive patterns of personality formation or what Elias called "psychic habitus." Different societies can engender very different kinds of people who exhibit distinctive patterns of psychological response, and in whom the process of childhood growth and development engenders distinctive patterns of cognition and rationality, i.e. different brains. What this implies is that modern liberal societies are almost certainly associated with a distinctive pattern of personality and even a "liberal brain" (Henrich et al. 2010). This is not to imply any normative notion that a "liberal" or "modern personality" is better – just different.

Modernization, disembedding, and the society of individuals

Modernity is, in this sense, the first form of society in which the questions "who am I?" and "what might I become?" have become even comprehensible for ordinary people. In a very real sense, modernity provides the first context in history in which the notion of personal freedom, that we now take for granted, is even thinkable. The chaotic maelstrom or urbanization combined with the rationalization of social relations and the waning sway of traditional cosmologies (disenchantment) produces people who are, for the first time in human development, experiencing themselves as distinct individuals with capabilities, desires, and subjective sensibilities that can find individual expression. Elias (1991) describes this process in terms of the changing I/we balance. The "society of individuals" is characterized by people whose image of themselves increasingly resembles what Elias refers to as the Cartesian "thinking statue" ("Cogito ego sum"); or the "closed person" (*Homo clausus*) separate, isolated, and autonomous, as opposed to the networked pluralities or figurations of interdependent "open people" (*Homines aperti*).

Although the ubiquitous perception of autonomy and individual agency in Western societies is validated through the narratives and procedures of every institution and social domain, the self is still fundamentally evoked through relations of interdependency with other selves in particular contexts (Mead 1934; Goffman 1959; Ross et al. 1991). Nevertheless, the process of individualization described in detail by Weber and many subsequent sociologists (e.g. Beck 1992: ch5) has certainly transformed all aspects of social life. Modern society is first and foremost, as Elias put it, a society of individuals (1991). The picture of mobile, more autonomous individuals with a greater capacity and necessity to make choices, is most certainly a central feature of the modern world.

This feature of modern life is so ubiquitous that it is hard to retain in mind its utter novelty in the long history of humanity. As Polanyi (1944, 1968) demonstrated so well, this epoch of individualization, driven by capitalist modernization, was a function of the waxing of the *market* and the *state*, and the waning of the "livelihood" domains of *house-holding and reciprocity* (see Quilley 2012). For Polanyi, the central dynamic during early modern capitalism was the disembedding of "economic" activity as a distinct domain, identifiable and separate from the wider cultural, religious, social, and political institutions of society. Whilst forms of market exchange are a feature of all recorded human societies, it is only with the appearance of the self-regulating market economy "directed by market prices and nothing but market prices" (ibid.: 45) that the process of provisioning and livelihood comes to be organized almost entirely around individual incentives for economic gain. It is only with such disembedded forms of economy that resources are able to flow freely with a view to maximizing such gains. Hitherto invisible and indivisible from other dimensions of culture and politics, the substantive matrices of group activity associated with the provisioning of communities become a separate, visible, and self-referential sphere, the domain of formal economics. With the emergence of the self-regulating market, "not blood tie, legal compulsion, religious obligation, fealty or magic [compel] participation in economic life, but specifically economic institutions such as private enterprise and the wage system" (Polanyi 1968: 81).

The details of this process are less important than Polanyi's insight that spatially and socially mobile individuals – agents capable of exercising instrumental rationality separate from the identifications of family and place and relating to/ identifying with the abstract survival units of state and market – could have emerged only in the wake of this process of the disembedding of the rural, land-based economy and the coercive processes that were required to consolidate the monopoly of state and market institutions as the dominant "survival units" (Elias 1991).

Individualism and the fibre of modern institutions

In addition to the "original sin" of agriculture (Sterlin, this volume), there are two necessary prerequisites for the emergence of liberal society. These are the violent disembedding of traditional forms of livelihood (Polanyi 2001) and the destruction of those tribal and place-bound "we-identities" that are incompatible with the emerging nation state (Elias 2010; Gellner 2008).

In modernizing societies, the individual – as a unit of analysis, of cognition and as a principle of organization – becomes so pervasive as to appear "natural." This is what Elias understood as our "second nature" or "habitus" – culturally acquired patterns of behaviour and cognition that are so deeply engrained as to become sufficiently automatic and pre-cognitive as to appear instinctive and an aspect of human nature (Elias 2012; Linklater and Mennell 2010: fn 74). Although projected as a universal feature of the human condition, this habitus only emerges in the context of complex, individuated state-societies in which (i.) individuals experience such spatial, occupational, and social mobility as to be significantly detached from place-bound kinship and tribally rooted survival units, and (ii.) that the state and market together function as a vehicle for welfare and security separate from the traditional constellations of kindred, community, and place.

Operating for the most part in the tramlines of Descartes, Kant, and Rawls, most Western political philosophy centres on a conception of agency and a type of agent that are only possible in the context of complex societies and high rates of economic growth, undergirded by a historically unprecedented throughput of energy and material. It is notable that commentators with wider time horizons and who do not take the metabolic foundation of consumer society for granted, are invariably drawn to older, often Aristotelian, Thomist, and non-Western philosophical traditions (Ophuls 2011; MacIntyre 1984).

A materialist account of ethics and agency: The triad of basic controls

Norbert Elias (1978, 1991) developed a materialist, developmental account of ethics and agency based on what he referred to as the "triad of basic controls." This deceptively simple framework was based on the empirical observation that controls in the domains or psychology, ecology, and society were mutually constituting, interdependent, and tended to develop in tandem. Thus, in the earliest period of human development, the emergence of a fire culture depended on the consolidation of a specific kind of "detour behaviour," i.e. psychological controls that allowed individuals to forgo immediate gratifications in expectation of future benefits (Goudsblom 1992). Fire-tending required people to resist the urge to put all the fuel on the fire at once, however cold it became – but to husband fuel resources to last over a period of time – a single night or a whole winter. Such *internal psychological restraint* was a capacity learned in the intergenerational process of socialization – and for children new to the function of fire-tending, this process involved *external social controls* which applied also to the division of labour associated with fire culture more generally: i.e. the expectation that if one person was collecting fuel, another might be foraging or preparing food. Simultaneously, human fire culture – the increasing frequency of anthropogenic fires in the landscape – had enormous evolutionary ecological impacts on the landscape, on the prevalence of other species and the development of ecosystems (Eisenberg 1998). Fire culture was therefore synonymous with *ecological controls*. In this way, it is possible to discern patterns in the weave of human development. Most fundamentally, there is a relationship between energy (*ecogenesis*) and social complexity (*sociogenesis*), and between the latter and processes of socialization and average personality formation (which means also the structure of the brain/ mind – *psycho-neuro-genesis*). In broad terms, the ecological impact of an energy regime (fire, agriculture, fossil fuels) is directly proportional to the level of social complexity, which in turn conditions processes of psychological development and the structure of average personality.

The conclusion of this perspective is that the biological universality of human nature is compatible with radical social-psychological variation (i.e. culture and personality); which is itself, at least in part, a function of particular ecological-energy regimes. People *really are different* in different societies/historical times, and this corresponds with the nature of both social complexity and the ecological signature. It further implies pretty strongly that any fundamental change in social complexity or the ecological signature of society – a change implied by the project of radical sustainability – would entail equally radical changes in the social relations and the character of the average personality.

Liberty and the epochs of human development

The idea of "liberty in the Anthropocene" immediately evokes a kind of thought experiment (see Table 15.1). Is it possible, by looking into the deep past, to identify the long-term ecological, sociological, and psychological modalities of liberty and individual agency, and, on this basis, develop some plausible ideas about what humanity might expect moving forward into a new and harsher eco-geological era?

Liberty in the Pleistocene (middle/upper Palaeolithic):

On the eve of the emergence of language and culture, human self-awareness and modes of cognition will have been very different to those of modern humans. This is true even if in genetic terms these people were more or less indistinguishable. Against Chomsky, and with Everett (2012, 2007), it seems that language is best understood (i.) as a highly specific technical innovation that is (ii.) culturally specific, (iii.) emerges in an individual brain only in that social-linguistic context, but (iv.) at the aggregate level of linguistic culture, significantly constrains and guides the modes of cognition and expression available within any culture. Thus, there is a mutually reinforcing relation between the neuro-cognitive development of particular brains and the complexity of symbolically mediated culture. A case in point is the contemporary 'stone age' culture of the Amazonian Piriha. Because their language doesn't contain subordinate clauses, numbers, or much by the way of tenses, there are forms of awareness about causality and relationships between people and objects through time that are very difficult to express. It seems to follow that, at the very least, the qualities of agency, subjective experience, and certainly the possible meaning of a concept such as "liberty" are likely to be so different as to be incommensurable and present insuperable translation problems. With regard to this incommensurability and radical untranslatability of different linguistic-cognitive worlds, it is not just that patterns of perception and lived and categorized experience are different (Elias 2007; Sapir 1949) but that, in a very real sense, both the objective world and its relation to human observers is different (Barfield 1965). Either way, at the very least, it seems certain that a complex idea like liberty can't really be projected backwards in any meaningful way. In such societies, the "we" dominates the "I." Daily life is coextensive with face-to-face interactions within a tight knit family or band grouping, which constitutes a primary survival grouping – primary in the sense that individual existence, safety, security, and procreation depend on membership of this group. There are no competing or alternative safety nets. As Elias (2007) pointed out, at this point in human development, the safety/danger ratio was very much weighted towards the latter.

The environment, though beneficent and giving, was also unpredictable. Humans competed with, hunted, and were hunted, by animals that were huge and dangerous. Even as language developed, high levels of danger and insecurity

TABLE 15.1 Epochs and the concept of liberty

	Pleistocene	Holocene	(later) Modernity	Anthropocene	
				High energy/successful eco-modernism	*Collapse*
Energy regime	Fire/ hunting-gathering	Solar/agrarian	Fossil fuel	Nuclear/high-tech solar?	Solar/agrarian
Controls over nature	Minimal	Low	High	Very high and rising	Lower/declining
Social complexity	Minuscule	Moderate	High and accelerating	Higher and accelerating?	Low/declining
Controls over/between individuals and groups	Minimal	Low	High	Higher/rising	Lower/declining
Social-psychological restraint/ *levels of interpersonal violence*	Low/Low	Low/High	Higher/Low	High/Low	Declining/*Rising*
Social-spatial mobility of individuals (as individuals)	Zero	Low	High and expanding	Higher and expanding	Low/declining

(*Continued*)

TABLE 15.1 (Continued)

	Pleistocene	Holocene	(later) Modernity	Anthropocene	
The I/We balance and survival units	We>I: *Family band*	We> I: *Clan/community*	I> We: *Individual/state/market*	I>We: *Market* Global governance and markets > nation state	We>I: *Re-nationalization/re-localization Nation-state and/or clan/community*
Liberty	Concept not applicable	Liberty for groups vis-à-vis other groups; and for high status individuals Agriculture: freedom from ecological constraints	Liberty for individual citizens, individuals (qua human rights) Freedom from ecological constraints (e.g. fossil fuel; the "green revolution") Transgender/transhumanism: incipient freedom from individual biological constraints	Liberty for individual citizens, individuals (qua human rights) Accelerating freedom from individual biological constraints Tensions between national citizenship and human rights qua supra-national/global institutions/entitlements	Contraction in the efficacy and meaning of liberty Re-emergence of biological and ecological constraints

Derived on the basis of Elias 1939, 1991, 1978; Quilley 2004, 2017, 2013.

and little capacity to build up food surpluses meant that there was little possibility to develop reality models of the natural world. Knowledge systems remained highly involved and imbued with high levels of fantasy. Barfield (1965) and Berman (2000) have both pointed to philological and anthropological evidence for a state of "original participation," characterized by a permeable interweaving of interior and exterior categories and phenomena that is now perhaps only fleetingly discernible in poetry. Certainly, although language carries with it the possibility of temporal awareness, this potential was to some extent unrealized (as examples such as the Piriha demonstrate – Everett 2012). People were almost certainly much more hodie-centric, immersed in the present and the flow of the moment, less able to plan future actions in relation to past events, and less able or inclined to situate a genealogical and coherent sense of self within the temporal flow of a life course. This very "present-centred" consciousness was reproduced in an oral culture the mechanics of which mitigated against complex abstract models of the world and of social relationships (Ong 1988). But by the same token, they were almost certainly not plagued by the neuroses and existential anxieties (e.g. the acute fear of death – Becker 1973), ontological insecurities (Giddens 1991), and narcissistic pathologies of self (Lasch 1978) that are pervasive in our own individualized and more temporally sophisticated societies. It seems likely also that a high degree of dependence on immediate kin combined with this paradoxical state of "original participation," might have mitigated against high levels of empathy for others, construed as "like us" and deserving of consideration by virtue of some abstract model of similitude (Rifkin 2009).

Liberty in the Holocene

The period of unusual climatic stability inaugurated by the Holocene saw an explosive spurt in human development. The pattern of surplus generation associated with agrarianization established a ratchet linking demographic growth, technical innovation, greater social complexity, and the expansion of agriculture as the dominant pattern of life. For the vast majority of people, social and spatial mobility were so constrained as to be unimaginable concepts. Survival units centred on family, community, and land and were usually hierarchical. Developing Barfield (1965), Morris Berman (2000) argues that the millennia of agricultural development were dominated by a very specific form of alienation – a sacred authority complex – that saw diffuse, "paradoxical," "participating consciousness" drain from the world and the increasing juxtaposition of a exanimate material world and a transcendent "spirit in the sky." Instantiated first by a pantheon of Indo-European sky gods, in time, such diversity was subsumed by the transcendent deity of the Abrahamic traditions. And although patriarchal and vengeful, Yahweh of the Old Testament did engender a latent possibility for a personal relationship between morally accountable individuals and the transcendent.

In this context, with literacy and greater complexity, civilizations periodically engendered a small degree of fluidity, albeit for very small groups of people – although the disembedding of the "I" from the "We" was tendential at best. But questions of personal liberty did become at least thinkable (Gilles 1987). In the classical world, the Judaic tradition, and eventually with the emergence of Christianity, the individual endowed with agency and responsibility emerges to take the stage in philosophical and theological thought experiments. And, at various times, such ideas bubble through into political movements with slave revolts and religiously inspired projects for the reformation of society (as with the coup that saw a subaltern Christian sect take over the Roman Empire).

Great societal variety notwithstanding, the common feature of traditional agrarian societies has been the extent to which imperial and state structures have emerged as a superstructure with little deep penetration into the lifeworld of ordinary peasants. In tribal societies such as Viking-era Iceland or fifth century Ireland, ostracism, banishment, or outlawing were tantamount to a death sentence. "We" affiliations were everything, and burned into the genealogical pattern of person and place-names. Heroic mythology and the greater complexity of society did engender the possibility for an experience or concept of liberty and freedom, and certainly the fact that people could be enslaved or more rarely liberated from such a state of bondage implies at least some understanding of individual freedom that might overlap with our own. But it was more usually the case that liberty was understood at the level of community, the integrity of society and its capacity to guarantee the process of intergenerational reproduction and succession. Mythological heroes were archetypes who responded to this need for societal integrity. The potential for an ontology of individual rights, obligations, and accountability, that was intrinsic to Judeo-Christianity, remained latent.

Liberty in modernity

Starting in the early modern period, rationalized capitalist agriculture, growing international trade and the exploitation of fossil fuels, lit the fuse for what would become the "great acceleration." Population growth, a rapidly expanding market economy and breakneck urbanization, created the conditions for the transition from the clannish and place-bound network of feudal fiefdoms to a nation market society of individuals – citizens, workers, and eventually consumers. This transition from (in Tönnies' language) "gemeinschaft" to " gesellschaft" involved an intrinsic co-relation between the nation-state and the market. The monopolies of violence, taxation, and education allowed the state to impose a uniform "high culture" across an increasingly homogeneous national culture – involving standardized, state-regulated exo-education systems, national currencies, national weights, and measures (Gellner 2008). And effective regulation provided the conditions for a flourishing economy and so consolidated the fiscal flows to the state as well as dividends for a growing bourgeoisie and upper middle class (Elias 2012). State and the market, though often in a relation of competition or tension, should be seen as a complex that has grown at the expense of informal, place-based, and family/clan-oriented forms of economic and social organization or livelihood.

All of the political and social liberties traditionally associated with modern societies depend absolutely on the vitality of this state/market complex and on the relative diminution of the domain of livelihood (see chapters by Mock and Quilley, this volume). Women's rights have been advanced on the basis of their status as individual citizens with equal rights under the law – not as members of families, partners in marriage, or people with attachments to this or that community or land. The same has been true of every other form of social emancipation – from sexual minorities to people with disabilities. Although the pattern of left/right politics has tended to pose the state and the market as antipodean antagonists, in fact they should be understood as a single complex whose dynamics establish the parameters for social democracy or liberal capitalism – however that is understood.

What this means is that the spatial and social mobility that is taken for granted in Western-type societies requires, as a prerequisite, a well-functioning market-state system which can function as the "survival unit" for free individuals. Thus, in the event of unemployment, sickness, or physical insecurity, individual well-being is guaranteed ultimately by access to the institutions of the state and/or the labour market. The price of a well-functioning political system and a dynamic and expanding market economy is that tribal and clan affiliations to place and community are dissolved in favour of abstract endowments consequent upon citizenship, and, more recently, the entitlements qua universal human rights.

Liberty in the Anthropocene

In his paper (this volume), Hensher describes a number of broad scenarios for global civilization in the Anthropocene – from "accelerationist" transhumanism (transcending the environment); eco-modernism (controlling); decoupling and green growth (accommodating with ease); voluntary degrowth (accommodating with greater constraint); and involuntary degrowth (collapse in the face of overwhelming constraint). I have previously developed a similarly comprehensive typology (Quilley 2017) as a basis for navigating the Anthropocene. But for the purpose of this essay, I want to concentrate on a distinction between (a) a form of eco-modernism defined by the maintenance and or expansion of existing levels of societal complexity, and (b) collapse defined by a significant loss of complexity, whether voluntary or involuntary, rapid (Kunstler 2005) or slow (Greer 2009).

Continuing societal complexity

In the case of *eco-modernism*, an accelerating trajectory of transcendence and control based on a high and expanding throughput of energy and materials would consolidate the emerging global society of individuals. The pattern of intensifying interdependence between individuals and groups would continue, further disembedding socially and spatially mobile individuals from kin and place-bound "communities of fate." Thus far, such developments have seen the consolidation of national societies based on social compacts associated with citizenship. And following Elias (2012, 2010), such developments have seen accelerated spurts of "civilization" characterized by the internalization of psychological constraints in relation to violence. In such circumstances, the realm of liberty, in the classic liberal sense of J.S. Mill, tends to expand – with individuals experiencing greater formal and substantive choices and fewer constraints. On the other hand, it is likely that such heightened individual agency is likely also to be accompanied by greater psychological repression, depression, unhappiness, and mental instabilities of various sorts (Lasch 1978; Becker 1973; Twenge and Campbell 2010). However, moving forward, a continuing trajectory of global integration would almost certainly engender a situation in which large numbers of affluent, culturally cosmopolitan "citizens of the world" become increasingly disembedded from and independent of, both national imaginaries and the nation-state as a principle survival unit (Quilley this volume).

Such a development would certainly create social and political tensions between such cosmopolitans and individuals affiliated to and dependent on national social compacts. The current tension between Leavers and Remainers in the UK Brexit debate is an incipient example of such tensions, which can also be detected very clearly in the emerging discursive juxtaposition of "cultural globalism" versus "nationalism." This can be understood, in part, as a tension between a survival unit defined by individual access to an increasingly global market on the one hand, or to a national citizen-based social compact on the other. The current incoherence of the radical left in the face of populist insurgencies is partly a function of their misapprehension of this tension – evident, for example, in the contrast between the open borders and international human rights championed by the hyper-liberal left on the one hand, and regulated labour markets, social insurance, and benefit entitlements in the context of a national welfare state, defended by traditional social democratic parties and increasingly by right-wing populists and protectionists, on the other.

Assuming that these tensions prove manageable, the scenario of accelerating eco-modernism would see the further movement towards hyper-liberalism and the rejection of all social and biological constraints on human choice and action. With respect to social constraints, this can be detected in the radical rejection of any links between individual rights and even minimal deontological commitments and obligations to the national society. The endgame for this kind of liberalism is, paradoxically, the rejection of citizenship per se, as an exclusive mask for in-group privilege. With respect to biological constraints, the now hegemonic insistence on the plasticity of human nature with respect to sex and gender and the burgeoning movement for "trans-rights" can be seen as a corollary to transhumanism more generally, and the Promethean impulse to deny any kind of natural constraint on human action.

A voluntary or involuntary loss of complexity

Any significant decline in the throughput of energy and materials would necessarily be associated with a significant decline in the complexity of society. The only way of mitigating this relationship would be technical innovations which have the effect of reducing the unit energetic cost of complexity. In Odum's terms this constitutes the "transformity cost" of civilization (Odum 2007; see Quilley 2017; Kish et al. 2016). The immediate impacts of such a scenario, whether voluntary or involuntary, would involve:

1. The re-embedding of economic activities and roles into the social, cultural, and religious pattern of life
2. A sharp decline in individual social and spatial mobility
3. (Initial) re-nationalization of economic life and the retrenchment around national social compacts as the primary survival unit
4. Fiscal crises resulting from a contracting economy would engender a legitimation crisis with regard to the national imaginary. Individuals would find themselves increasingly dependent on new primary survival units organized around kindred, community, occupational associations, and place. This unravelling of the existing social compact would certainly be experienced as a political shock and present an existential threat to the security of many citizens. But it would also expand the space for the innovation of completely new or the rehabilitation of older survival units (Quilley 2012; Zywert and Quilley 2018)
5. Over a longer period, the declining state monopolies with respect to taxation and violence would engender changes in the average personality structure, decreasing patterns of interdependence between individuals and groups, and what amounts, in Eliasian terminology, to a process of 'decivilization' (Mennell 1990)
6. Greater ascription in social and occupational roles would almost certainly be accompanied by greater ascription in gender roles with the re-emergence of a strong sexual division of labour. Drivers of this change would involve variously: the retrenchment of legal and civil infrastructure promoting equal rights; the declining availability of birth control; the reduction or loss of state-funded childcare services; a radical reduction in occupational complexity and the loss of many service sector and care occupations from the formal economy; the decline or complete loss of education institutions; and in an energy-poor world, a heightened premium on physical bodily strength in many occupations currently dependent on machine labour

The philosophical concept of liberty that dominates the Western imaginary is predicated on the idea of socially and spatially mobile individuals. By any yardstick, any significant loss of complexity would necessarily reduce such mobility. Whether voluntary or involuntary, such a scenario would involve a loss of liberty and an enormous reduction in the choices available to individuals.

Discussion: Liberty in the Anthropocene

Can the concepts of liberty characteristic of modernity (however inadequately realized) survive into the deep future? Or was modernity a fossil-fuelled blip in human history to be followed by some kind of relapse? What can be said with any certainty?

First of all, the modern, liberal idea of liberty depends absolutely upon the modern society in which it emerged. It is a high energy, high social complexity idea that makes sense only when a combination of state institutions and effectively functioning markets engender high levels of spatial and social mobility, and allow individuals to become detached from traditional familial and community-based survival units. Liberty in this sense is as incongruous and irrelevant to the face-to-face societies in which people lived for most of the last million years, as Barfield's (1965) cognitive–affective state of "original participation" would be if applied to modern global society.

Second, ultimately, this social complexity depends on a low-entropy energy regime (Odum 2007; Quilley 2011, 2012). Without a continuing and rising throughput of energy, the growth economy will collapse, undermining the regulatory and fiscal capacities of the state, and contracting the scale and scope of disembedded price-setting markets. Any such decline would automatically undermine the state-market's capacity to function as a survival unit of last resort. And as a result, family, clan, and community/placed-based forms of mutual obligation and support would necessarily re-emerge to take up the slack and provide very basic forms of security.

Third, what this means is that the continuing existence of recognizable discourses and institutions of liberty is inextricably tied up with the ecological-economic viability of the global industrial economy. It may be possible intellectually to re-define liberty in such a way as to avoid this problem (not least in this volume). However, I have taken the view that a productive debate is only really possible on the basis of a shared vernacular understanding of the concept that derives from the perceptions and intuitions of individual agents operating in a liberal society of individuals. Eco-modernists such as Michael Shellenberger and Stewart Brand (Asafu-adjaye, J. et al. 2015) can afford to be optimistic as to the prospects of liberty. On the other hand, eco-pessimists who see some kind of contraction or "degrowth" as an unavoidable concomitant of Anthropocene economics have legitimate cause for alarm (Quilley 2017, 2013). "Liberty" is only as viable as a modern economy that can deliver social-spatial mobility and a continuing flow of resources to fund a plethora of state institutions (democracy, the legal system, education, health, policing, military, transport and communications infrastructure, etc.).

Fourth, the implied paradigm shifts are not necessarily comparable. Structures of meaning and patterns of thought change fundamentally between societies depending on the I/We balance, the extent of individualization, degrees of spatial/social mobility, level of development of the state – and also ontological frameworks (compare the immersive "participating consciousness" of Palaeolithic hunter-gatherers in which all social relations and ecological relations are mediated by reciprocity and gift exchange – with say the contractual, hyper mobile, highly individualized consumer societies). It is difficult to say with certainty whether "more" or "less" liberty is a meaningful concept or whether more/less could be evaluated objectively as better/worse. Having said this, it does seem almost certain that people whose personalities and value-systems have been moulded in the context of modern societies would experience any sharp "loss" of liberty (social/spatial mobility, judicial freedoms, etc.) very negatively.

Conclusion

A primary conclusion from this analysis is that intellectual conceptions and subjective experiences of whatever we might understand by liberty depend to a very great extent on the technical, societal, and ecological context – and specifically on the level of societal complexity. Hensher's recommendation (this volume), to maximize optionality, is a reasonable response to a situation of paradigmatic uncertainty and risk. Certainly, it is associated with an intrinsic bias – that modern, liberal conceptions and subjective perceptions are inherently more valuable than the unknowable ontological and experiential outlook of, say, Paleolithic hunter-gatherers. What Barfield refers to as "participating consciousness" is essentially completely incompatible with any modern conception of liberty, including all of those represented in this volume. However, Hensher is also right to warn against the conflation of technical options with ideological worldviews. The observation that nearly all extant visions of feminism depend on high levels of economic and social complexity, which in turn may prove incompatible with ecological integrity (Quilley this volume), is precisely that – i.e. an observation and not a normative or ideological commitment. On the other hand, embracing this distinction and engaging in the widest possible process of ideological miscegenation and dialogue is probably the most effective way to engender ideational novelty and to make accessible the greatest array of options in the "adjacent possible" (Kauffman 1995). This is the intent of my exploration of the political-economic and ideological/normative links between the state, market, and livelihood (Quilley this volume). Although this would seem to constitute a practical strategy of pre-figuration, it should be recognized that it relates only to the immediate constraints of the "near bad future." With regard to the longer geological time frames evoked by the Anthropocene, the very idea of human agency seems to smack of hubris. Seen against the time horizons and energy-complexity imperatives of the "plus longue durée," the poet Robinson Jeffers is probably realistic in qualifying the efficacy and autonomy of human agency. The beauty of modern civilization, he argues, is "not in the persons but in the disastrous rhythm ... the dance of the dream-led masses down the dark mountain" (Rearmament 1935, quoted in Kingsnorth and Hine 2010).

References

Asafu-adjaye, J. et al. (2015). *An Ecomodernist Manifesto*. http://static1.squarespace.com/static/5515d9f9e4b04d5c3198b7bb/t/552d37bbe4b07a7dd69fcdbb/1429026747046/An+Ecomodernist+Manifesto.pdf

Barfield, O. (1965). *Saving the Appearances; A Study in Idolatry*. New York: Harcourt, Brace Jovanovich.

Berlin, I. (2002 [1969]). "Two concepts of liberty." In I. Berlin, (ed.), *Four Essays on Liberty*, London: Oxford University Press. New ed.

Beck, U. (1992). *Risk Society*. Cambridge: Polity.

Becker, E. (2014 [1973]). *Denial of Death*. London: Souvenir Press.

Benedict, R. (1934). *Patterns of Culture*. Boston: Houghton Mifflin.

Bentham, J. (1823). *An Introduction to the Principles of Morals and Legislation*. London: Printed for W. Pickering.

Berlin, I. (1969). "Two concepts of liberty." In *Four Essays on Liberty*, Oxford: Oxford University Press, 118–172.

Berman, M. (2000). *Wandering God: A Study in Nomadic Spirituality*. SUNY Press.

Beveridge, W. (1960 [1944]). *Full Employment in a Free Society: A Report* (2nd ed.). London: Allen & Unwin.

Eisenberg, E. (1998). *The Ecology of Eden: An Inquiry into the Dream of Paradise and a New Vision of Our Role in Nature*. New York: Vintage.

Elias, N. (2007 [1987]). *Involvement and Detachment*. 2nd Revised Edition. Vol.8. Complete Works of Norbert Elias. Ed. Stephen Quilley. Dublin: UCD Press.

Elias, N. (2010 [1991]). *The Society of Individuals*, edited by Robert van Krieken, Dublin: UCD Press.

Elias, N. (2012 [1939]). *On the Process of Civilisation* [note new title], edited by Stephen Mennell, Eric Dunning, Johan Goudsblom and Richard Kilminster, Dublin: UCD Press.

Elias, N. (2012 [1978]). *What is Sociology?* edited by Artur Bogner, Katie Liston and Stephen Mennell, Dublin: UCD Press.

Everett, D. (2007). 'Challenging Chomskyan Linguistics: The Case of Pirahã,' *Human Development*, 50(6): 297–299.

Everett, D. (2012). *Language: The Cultural Tool* (1st ed.). New York: Pantheon Books.

Frank, M. C., Everett, D. L., Fedorenko, E., & Gibson, E. (2008). "Number as a cognitive technology: Evidence from Pirahã language and cognition." *Cognition*, 108(3); 819–824.

Freeden, M. (1986 [1978]). *The New Liberalism: An Ideology of Social Reform*. Oxford: Clarendon Press.

Gaus, G. F. (1996). *Justificatory Liberalism: An Essay on Epistemology and Political Theory*. New York: Oxford University Press.

Gellner, E. (2008). *Nations and Nationalism* (2nd ed.). Ithaca, NY: Cornell University Press.

Giddens, A. (1991). *Modernity and Self-identity: Self and Society in the Late Modern Age*. Cambridge: Polity Press.

Gilles, A. (1987). *The Evolution of Philosophy: An Overview of Western Thought as it Relates to Judeo-Christian Tradition*. New York: Alba House.

Goffman, E. (1959). *The Presentation of Life in Everyday Life*. Penguin.

Goudslom, G. (1992). *Fire and Civilization*. London: Allen Lane.

Green, T. H. (1986 [1895]). *Lectures on the Principles of Political Obligation and Other Essays*. Paul Harris and John Morrow (eds.), Cambridge: Cambridge University Press.

Greer, J. (2009). *The Ecotechnic Future: Envisioning a Post-peak World.* Gabriola Island, BC: New Society.

Hayek, F. (1961). *The Constitution of Liberty.* Chicago: University of Chicago Press.

Henrich, J., Heine, S., & Noransayan, A. (2010). "The weirdest people in the world." *Behavioural and Brain Sciences,* Page 1 of 75, doi:10.1017/S0140525X0999152X.

Hofstede & McCrae (2004). "Personality and culture revisited." *Cross-Cultural Research,* 38(1): 52–88.

Kluckhohn, C., & Murray, H. (1953). *Personality in Nature, Society and Culture.* New York: Alfred A. Knopf.

Kauffman, S. (1995). *At Home in the Universe: The Search for Laws of Self-organization and Complexity.* New York: Oxford University Press.

Keynes, J. (2007). *The General Theory of Employment, Interest and Money* (New ed.). Basingstoke [England]; New York: Palgrave Macmillan for the Royal Economic Society.

Kingsnorth, P., & Hine, D. (2010). *The Dark Mountain Manifesto Dark Mountain Project.* Retrieved from https://dark-mountain.net/about/manifesto/

Kish, K., Hawreliak, J., & Quilley, S. (2016). "Finding an alternate route: Towards open, eco-cyclical, and distributed production." *Journal of Peer Production,* 9 (September). http://peerproduction.net/issues/issue-9-alternative-internets/peer-reviewed-pap ers/finding-an-alternate-route-towards-open-eco-cyclical-and-distributed-production/

Kunstler, J. (2005). *The Long Emergency: Surviving the End of Oil, Climate Change, and Other Converging Catastrophes of the Twenty-first Century.* New York, Berkley: Grove Press, Distributed by Group West.

Lasch, C. (1978). *The Culture of Narcissism: American Life in an Age of Diminishing Expectations.* New York: Norton.

Lindholm, C. (2001). *Culture and Identity: The History, Theory, and Practice of Psychological Anthropology.* New York: McGraw-Hill.

Levine, R. A. (2001). "Culture and personality studies 1918–1960." *Journal of Personality,* 69(6): 803–818.

Linklater, A., & Mennell, S. (2010). "Norbert Elias, the civilizing process: sociogenetic and psychogenetic investigations—An overview and assessment." *History and Theory,* 49(3): 384–411.

Locke, J. (1960 [1689]). *The Second Treatise of Government in Two Treatises of Government,* Peter Laslett, ed. Cambridge: Cambridge University Press, 283–446.

MacIntyre, A. (1984). *After Virtue: A Study in Moral Theory* (2nd ed.). Notre Dame, IN: University of Notre Dame Press.

Mead, G. H. (1934). *Mind, Self, and Society.* C. W. Morris (ed.), University of Chicago Press.

Mead, M. (1935). *Sex and Temperament in Three Primitive Societies.* London: Routledge.

Mennell, S. (1990). "Decivilizing processes – theoretical significance and some lines of research." *International Sociology,* 5(2): 205–223.

Mill, J. S. (1963). *Collected Works of John Stuart Mill,* J. M. Robson (ed.), Toronto: University of Toronto Press.

Odum, H. (2007). *Environment, Power, and Society for the Twenty-first Century: The Hierarchy of Energy.* New York: Columbia University Press.

Ong, W. (1988). *Orality and Literacy.* London: Routledge.

Ophuls, W. (2011). *Plato's Revenge: Politics in the Age of Ecology.* MIT Press.

Polanyi, K. (1968). *Primitive, Archaic, and Modern Economies: Essays of Karl Polanyi.* G. Dalton, (ed.) Garden City, NY: Anchor Books.

Polanyi, K. (2001 [1944]). *The Great Transformation: The Political and Economic Origins of Our Time* (2nd ed.) Boston: Beacon Press.

Quilley, S. (2004). "Ecology, 'human nature' and civilising processes: Biology and sociology in the work of Norbert Elias." In S. Loyal & S. Quilley (eds.), *The Sociology of Norbert Elias*, Cambridge: Cambridge University Press, pp. 42–58.

Quilley, S. (2011). "Entropy, the anthroposphere and the ecology of civilization: An essay on the problem of 'liberalism in one village' in the long view." *The Sociological Review*, *59*(June): 65–90. doi:10.1111/j.1467-954X.2011.01979.x.

Quilley, S. (2012). "System innovation and a new 'great transformation': Re-Embedding economic life in the context of 'de-growth.'" *Journal of Social Entrepreneurship*, *3*(2): 206–229.

Quilley, S. (2013). "De-growth is not a liberal agenda: Relocalisation and the limits to low energy cosmopolitanism." *Environmental Values*, *22*(2): 261–285.

Quilley, S. (2017). "20 navigating the anthropocene: Environmental politics and complexity in an era of limits." *Handbook on Growth and Sustainability*, 439.

Rawls, J. (1999). *A Theory of Justice*, revised edition. Cambridge, MA: Harvard University Press.

Rifkin, J. (2009). *The Empathic Civilization: The Race to Global Consciousness in a World in Crisis*. New York: J.P. Tarcher/Penguin.

Robbins, L. (1961). *The Theory of Economic Policy in English Classical Political Economy*. London: Macmillan.

Ross, L., Nisbett, R., Gladwell, M. (1991). *The Person and the Situation: Perspectives in Social Psychology*. Pinter and Martin.

Sapir, E. (1949). *Culture, Language, and Personality*. Berkeley: University of California.

Spencer, H. (1851). *Social Statics: Or, The Conditions Essential to Human Happiness Specified, and the First of Them Developed*. London: J. Chapman.

Steiner, H. (1994). *An Essay on Rights*. Oxford: Basil Blackwell.

Twenge, J., & Campbell, W. (2010). *The Narcissism Epidemic: Living in the age of entitlement* (1st Free Press trade pbk. ed.). New York: Free Press.

Zywert, K. and Quilley, S. (Eds) (2018). *Health in the Anthropocene: Living well on a Finite Planet*. Toronto, Canada: Toronto University Press.

16

FOREST ON TRIAL

Towards a relational theory of legal agency for transitions into the Ecozoic[1]

Iván Darío Vargas Roncancio

Introduction: Vegetal agencies in the law

With the serene joy of decades of experience and struggle, Don D, an indigenous elder from the Cofán community in Bajo Putumayo, Colombian Amazon, was seated on an old-looking stool holding a piece of cord. "P passed away," he said, and then remained silent. As I introduced myself and paid my respects, I recognized the skin of the manioc plant in the elder's hand. "*El abuelo es yuca*" ("the elder is a manioc plant"), I thought, as I mused about plants as persons in Amazonia (Gagliano, 2013). Are human engagements with other beings limited to the dubious epiphany that we all depend on the mineral, vegetal, and animal life of the world? What do other-than-human beings have to do with the law in this region?

National legislations and governance models across the world increasingly recognize the legal subjectivity of other-than-human beings (Acosta and Martínez, 2011; Harris, 2014; Youatt, 2017). The contested clause of the rights of nature (RN),[2] for instance, is a growing response to economic practices underpinning the "inter-related global crises of climate, food, energy, poverty, and meaning" (Escobar, 2016, p13).

While the RN express the radical interdependence between natural and social systems, dominant environmental governance models in Amazonia seem deeply entangled with what Colombian legal scholar Gregorio Mesa calls an "ethics of consumption." He refers to the social mindsets and institutions casting nature as an endless quarry of material goods and ecosystems services to meet ever-expanding human needs (2008, p333).

Increasing eco-centric legal proposals such as the RN emerge in the midst of neo-extractivist/colonial practices with lasting negative impacts on socio-ecological systems (Gudynas, 2009). This tension is an expression of the pervasive ontology of separation between nature and culture in much of environmental law and governance models today (Atleo, 2011; Vargas et al., 2019). How can a new legal ontology contribute to a shifting paradigm: from a highly regulatory environmental approach to a relational and knowledge-grounded ecological jurisprudence? (Burdon, 2012; Cullinan, 2011; Pelizzon, 2014).

By drawing connections between the law and larger socio-ecological systems (Garver, 2013, 2019), the first section of this chapter challenges a standard definition of the law as a system of norms while rendering visible the ontological and cognitive dimensions of legal practices (Vermeylen, 2017; Philippopoulos-Mihalopoulos, 2017; Winter, 2001).

In particular, it analyzes how the notion of personhood (see Anker, 2017) – central to Western theories of rights and justice (López, 2018 [2004]) – conceals understandings of the law as a potential emancipatory and world-making tool for transitions into the Ecozoic, an era of mutually enhancing human–Earth relationships (Berry, 1999; Berry and Swimme, 1992).

Based on ethnographic encounters in the Colombian Amazon, as well as a short review in the field of plant communication and intelligence (Gagliano et al., 2017; Mancuso and Viola, 2015; Marder, 2013, 2016; Myers, 2015), the second section illustrates how non-human collectives, I tentatively call non-persons, overflow the ontological stability of objects and subjects. This working notion offers analytical keys for a concept of legal agency beyond the modern divide actualized by the idea of personhood.

While much of current debates about the RN are framed in terms of granting legal personhood to non-humans (Youatt, 2017), I suggest calling attention to the relational framework of indigenous thought foregrounding the earthly co-emergence of humans and other-than-humans in legal systems (Borrows, 2016; Kimmerer, 2013). Inspired by the principle of interdependence (Atleo, 2011), the chapter concludes by asserting how other-than-human legalities might become compelling sources of a legal ontology beyond human-only, hyper-regulatory, and interculturality-blind environmental legal frameworks today, as well as the questions this effort raises.

Why plants?

I consider the case of plants, among others, for two main reasons. First, much attention has been given to the figure of the animal in social theory (Few and Tortorici, 2013), while studies on the relationship between plants and social systems have received much less attention (Marder, 2013; Gagliano, 2013; Gagliano et al., 2017). Moreover, the sessile character of plants affords a compelling case against an idea of agency based on human perceived movement as an index of change.

An increasing number of studies ranging from plant communication and intelligence (Trewavas, 2016) to vegetal neuro-physiology (Mancuso and Viola, 2015) and plant bio-acoustics, reveal a vegetal sensorium functionally similar to organisms with centralized nervous systems (Gagliano et al., 2018). As sign-making selves, for example, some plants can locate water sources through their roots by sensing the sound vibrations of water moving inside pipelines.

The ability of plants to detect vibrations "may represent a very efficient way of capturing information from distant sound sources for orientation towards water" (Gagliano et al., 2017, p152). A suggestive proposition, the idea that plants can *listen* and orient themselves towards water sources might well sound like a mere humanization of biological phenomena. Yet, as anthropologist E. Viveiros de Castro would argue, "when everything is human, the human is an entirely different thing" (Viveiros de Castro, 2014).

Moreover, Myers develops an ethnographic argument on the limits of the anthropogenic classification of plant life, namely, the tendency to define plants' attributes in terms of human attributes. She uses expressions such as the "plant turn" (2015, p40) or "vegetal epistemology" (p42) to account for a non-anthropocentric stance to plant modes of knowledge and being.

However, a focus on relations, encounters, and "mutual differences" (de la Cadena, 2015) may redefine much-praised human attributes such as intelligence and memory as emergent properties of inter-being encounters (Haraway, 2008). They are not specific capacities of humans (i.e. the human point of view) or plants (i.e. the plant point of view), defined as discrete entities themselves (Escobar, 2018). To be sure, plants and humans co-emerge in symbiosis in legal worlds (Margulis and Sagan, 2002).

Personhood reinforces the ontological dualism between humans and nature. Can we think of legal agency beyond this divide?

In the foreword to *Lively Legalities*, Wolfe asks his readers the following compelling question:

> What sense does it make that a highly developed animal such as a tiger or an orca is regarded as 'the same' as a toaster or a pile of bricks, while in US law at least, corporations and ships of state are legally designated as persons?
>
> *(Wolfe, 2016, pxiv)*

Indeed, natural and juridical persons such as humans and corporations equally have some legal standing, such as the ability to seek redress before a court of law (Stone, 1972). While the distinction between the natural (i.e. laws of "nature") and the juridical (i.e. positive norms) is not a settled discussion, the fact that legal entities such as corporations can be considered persons intriguingly expands the notion of personhood beyond the human.

For Dewey, the "'person' signifies what the law makes it signify," and it is therefore an empty and positional signifier (1926, p655). Similarly, Marder elaborates on the neighbouring notion of subjectivity as the capacity to act with intent to "actively shape the world" (2016, p56). For example, organisms such as plants can actively shape their own milieus, thus expressing a subjective or person-like property akin to the notion of rights.[3]

Does reckoning the legal personhood to non-humans such as plants and forests reinstate the ontology of separation between humans and the rest of life at the root of dominant legal systems? An almost commonsensical perspective would say that to be a person entails "enjoying [...] rights in and of themselves" (Esposito, 2010, p121). In the field of legal theory, however, it remains unclear whether the idea of rights – including the right to obtain legal remedy – can exist without the modern concept of the person.

For Esposito, "the enclosed space of the person" is not a necessary condition for legal rights (2010, p122); in fact, "no one is born a person [...] Some might become a person, but precisely by pushing those that surround him into the dimension of the thing" (2010, p126).

Pushed into the dimension of thing-like phenomena, entities and processes such as water systems, forests, and nitrogen cycles are devoid of agency (and rights) unless they are made into persons through the law. In other words, non-persons (or things) would be, at best, objects of environmental protection with no rights "in and of themselves."

Thus, the ability to have rights and duties expresses the inherent legal capacity of persons as opposed to the inherent natural behaviour of things. Dewey furthers this ontology of separation when affirming that "molecules or trees" behave "exactly as they do" by the force of nature – rather than by the rule of law (1926, p661). As he allocates agency on the side of the person at the expense of the passive materiality of non-persons ("molecules and trees"), Dewey actualizes the pervasive distinction between subjects and objects via a theory of personhood.

Similarly, Stephens suggests a compelling tripartite spectrum to define nature: from "*relatively* untransformed nature at one pole, borderline places such as traditional farms and country paths [...] and the world of completed artefacts and radically instrumentally transformed goods [...] the other pole" (2019). While this notion differentiates the intensity and scale of human intervention over an *external* nature, it seems to locate agency and liberty on the human-end exclusively, thus re-inscribing the modern tension between subjects and objects just described.

To state it in terms of "geologian" Thomas Berry, the radical discontinuity between human and other-than-human beings (aka persons and non-persons), and the ensuing "bestowal of all rights on the humans" alone, underpins the ecological and cultural devastation of the planet (1999, p4). The task ahead is then "reinventing the human" by re-embedding social institutions such as the law within the broader community of life. I see this project connected to the re-evaluation of deeply entrenched notions such as the pair subjectivity/personhood.

Situating non-personhood within legal ontologies

While some scholars contest the idea of rights given the little evidence of their real effectiveness (Posner, 2014), legal theory might benefit from a conception of agency beyond the legal twins of personality/subjectivity, thus expanding the notion of rights. This endeavour, to be sure, re-conceptualizes the limits of a legal ontology based on the separation between persons and things; the disentanglement of legal norms and vital relations; and the decoupling of social and ecological justice.

A relational and experience-grounded approach to the law, on the other hand, seeks to de-centre the human in legal systems (Grear, 2017) to acknowledge the recursive interactions between humans and other-than-human selves (Kohn, 2013). This relational idiom privileges the integrity of the whole Earth community in the long term over the interests of humans alone (Cullinan, 2011). Yet, what does the expression "legal ontology" refer to in this context?

Broadly speaking, the law deals with a universe of relations between individuals, communities (human and not), states, and "elementary groupings themselves" (Graham, 2011, p15 in Burdon, 2010, p28). However, usually depicted in terms of norms and procedures separated from the larger socio-ecological processes they regulate, Western legal theory tends to overlook the relational and material dimensions of law-making. It takes for granted notions of personhood and subjectivity profoundly tied to mechanistic/Cartesian worldviews (Anker, 2017; Capra and Mattei, 2015).

Under this predominant mechanistic paradigm, the law is conceptualized as an autonomous human institution separated from its socio-material processes (Burdon, 2012; Braverman, 2016). Rather than something circumscribed to a set of abstract norms separated from the concrete sensimotor handling of living organisms (Varela, 1999), the rules-we-live-by arise in recursive engagement with the world humans co-create with other selves. Thus, the law does not denote something pre-given to the mind, but "something we (humans and not) engage in by moving, touching, breathing and eating" (Varela, 1999, p7).

In contrast, an ontological approach to the law considers the world-making potentials of legal systems, and not only its prescriptive attributes. Beyond the universalistic formalism of top-down legal codes (Winter, 2001), legal norms are grounded or embedded in the concrete experience of humans, and I will later argue, other-than-humans. A systems-based legal ontology thus brings to trial mechanistic and deterministic approaches to the law.

On how ontological choices may determine legal frameworks

As early as 1972, US Justice William O. Douglas argued that trees should have legal standing. For him,

> [i]nanimate objects are sometimes parties in litigation. A ship has a legal personality, a fiction found useful for maritime purposes [...] So it should be as respects valleys, alpine meadows, rivers, lakes, estuaries, beaches, ridges, groves of trees, swampland, or even air that **feels** the destructive pressures of modern technology and modern life.[4]

To the extent that the entities in this list should be granted rights, can we say that they should be considered persons as well? What ontological commitments do we make when personifying "nature"?

More than enclosed entities, forests, meadows, swamplands, and the like are often defined as bundles of ever-renewing relations eluding the stability of the bounded-self (Haraway, 2008). In order to have standing, however, these entities require the legal quality of personhood regardless of their living and human-like attributes. To be sure, the entity in question ought to be a person first (it needs to occupy the "enclosed space" of the bounded self) in order to have rights.

To be a person, however, one does not necessarily need to be alive. For example, a corporation can become a person in order to hold legal standing and rights. This decoupling between "personhood" and "life" creates fictional persons (corporations) with rights regardless of ecological considerations.

This ontology of separation, again, divorces the law from larger socio-ecological processes by affording legal standing to the person, and thus re-inscribing the rift between humans and other-than-humans. As mentioned before, a "'person' signifies what the law makes it signify" (Dewey 1926, p655) or, in other words, the law is made into an autonomous sphere of action fictionally – but quite productively – separated from the life systems it depends upon.

Fluid relations with elusive embodiments, non-persons such as water cycles, soils, and forests, best describe what Amazonian indigenous ethno-botanist Abel Rodríguez calls the "figure of life" – or how life emerges and takes hold (Rodríguez, 2014). Extending this argument to the legal realm, non-persons could claim a strong presence within the law that includes the language of rights, but is not reducible to it. Rather than bearers of rights afforded by humans, these relations are sources of the law (Borrows, 2016), as well as law-making agents in their own right (Strathern, 2005). In this context, to *create the law* stands for a quality of what I provisionally call semiotic relations, namely, the relational and meaning-producing character of life (Kohn, 2013).[5]

Standing on the legal stage, the actor conceals the human and its constitutive relations underneath the mask of the person

As the idea of corporate legal personality illustrates (Dewey, 1926), the realm of person-like phenomena is rather stable, predictable, and limited. The realm of relational-like phenomena, on the other hand, appears rather vast, unpredictable, and politically contentious. What happens with obliterated life relations unaccounted for by the law as *persons*?

In the multiple and relentlessly productive field of Amerindian cosmologies, humans, animals, plants, and spirits conceal an internal human form (Descola, 2013; Viveiros de Castro, 1998, 2014). The person stands for a mask or clothing (Rodriguez's "figure of life") concealing underneath human-like properties of non-human selves such as rivers, plants, and forests (Kohn, 2013).

While living beings can only become juridical persons through an act of legal recognition reinstating the separation between the "person" and the "thing" (Esposito, 2010), non-personality re-organizes the relational web of legal agencies. Here is an example.

An 80,000-year-old organism, the *pando* (*Populus tremuloides*) looks like a grove of separated trees with a densely intertwined complex root system underneath the topsoil.[6] As legal theory increasingly grounds its conceptual toolkit within the life sciences, non-persons such as the *pando* pose uncanny challenges to a theory of legal agency based on the twin notions of subjectivity and personhood. Especially when we learn that more than half of the biomass of this life system reverberates under our feet, and beyond the human purview (Wohlleben, 2016, p85). The notion of personhood here reveals itself inadequate, for it manages the elusive stability of life relations through the counting of persons-for-the-law.

The legal person-making machine obscures what stands behind the legal mask of personhood: a messy and often unpredictable "meshwork" of relations escaping the logic of the enclosing law (Ingold, 2011). In other words, while the notion of personhood is a legally produced quality of humans and other-than-human beings, this notion renders invisible the relations that make life entities emerge from underneath the field of human visibility. I am suggesting that life relations often shape the legal purview of personhood and thus require concepts attuned to its material unfolding.

Deemed a necessary legal fiction, the notion of personhood can certainly be extended to highly heterogeneous kinds of entities from corporations and states to forests and rivers. The list is endless. My argument advocates for a legal ontology that considers how life works, and what kind of legal theory emerges from this realization. I will explain this point further below.

In a context of increasing recognition of the rights of nature (Gudynas, 2009; Burdon, 2010), one may wonder whether legal categories used to adjudicate human affairs can easily be transferred over to non-human beings. Particularly in the case of interconnected entities seen as discrete selves above ground (the *pando*), while concealing underneath a complex organization beyond the reach of direct human perception (the law).

Legal adjudication of personhood meets not only the constraints of human perception – and their specific modes of symbolic representation – but also the unexpected materiality of inter-being encounters reverberating under our feet. Before any juridical personalization, Esposito insists, living beings constitute a plane of indivisibility whereby life (human and not) *is* to itself; "in which the form, precisely of life, is the form of its own content" (2010, p132). The mask (the person) and the actor (life relations) become one and the same.

Up to this point, I have suggested that life entities such as plants and forests can be persons either epistemologically separated from their vital relations or fully placed within them. Shifting our dominant Western theories of rights necessitates an opening up of the enclosed space of the person in order to consider the relations behind the legal fiction (or mask) of personhood. It is the law that needs to be integrated into the logic of life, and not the other way around.

Partially guided by the threads of ethnography and plant science, the next section offers some empirical grounding for the larger argument on the relational (or non-personal) ontologies of the law. I ask how the legal adjudication of rights may look from the vantage point of plants and forests as relations rather than (bounded) substances.

Indigenous law, plants, and ritual: A story of Amazonian legal concepts[7]

"*De **Aquí** viene el derecho indígena*" ("indigenous law comes from *here*") said the *taita*[8] in the midst of an intense vertigo with the *ayahuasca* (*Banisteriopsis caapi*) brew[9] known as *chuma*.[10] Like a firefly hovering in the background of my memories, this enigmatic statement has stayed with me for quite some time.

Some time ago, a friend invited me to the Lower (*Bajo*) Putumayo region located some miles north of the Colombian border with Ecuador. "There is a ceremony with the Cofán elders, and a group of visitors from the Cumbal region (Colombian Andes)," he said. "What is a *toma*?" I asked, halfway between curious and bewildered. "*Ayahuasca*! But this time it's going to be with the *mayores* (*elders*). They know how to manage this plant."

I had heard quite a few stories of people having some intense experiences with the "soul creeper vine," also known as *yagé* (Schultes and Hofmann, 2012). But I had also heard one must keep a strict diet before ingesting the brew with a "person who knows it," as my friend said, and in the appropriate ritual setting. With trepidation, I accepted his invitation, remaining highly undecided about doing the *toma* itself. Such a bidding, however, had to be endorsed by a middle-aged *Ingano taita* from the Upper Putumayo region, who was summoned to lead the ritual.

Once he had decided that I had rightful intentions, the *taita* accepted me. The plant "calls upon the people needing it," he said, and I imagined a bunch of entangled systems of underground roots connected to an ancient stump somewhere in the Amazon, sending chemical signals all the way from the canopy to the Andean mountains. Soon, we would set out on the road to the unexpected, yet open to dialogue with the "sacred vine" on its own terms (Schultes and Hofmann, 2012).

We met at the bus terminal. A good *Ingano* seasoned in the arts of trading plants and ritual objects all over the country, the *taita* bargained with an office clerk to get a reduced fare on the tickets; we bought some snacks for the trip, and finally set forth on the way down to the *Bajo*. After around 12 hours of paved roads and hilly scenery, we arrived in Mocoa, capital of Putumayo and the urban hinge between the Andes and the Amazon.

Founded by Avendaño in the middle of the 16th century, this city vanished within a few decades of its creation. Refusing to abide by the sword, the cross, and the Bible, a group of audacious locals, known as the *Andaquíes*, burned the city down, and then returned to the forest, never to leave it again. With the spectre of the *Andaquíes* quietly hiding behind the bushes, we arrived at our final destination a few hours later.[11]

"Light up a tobacco before getting in. It will help you to *pensar bonito* ["to have beautiful thoughts"]," the *taita* said as we entered the *resguardo*. "This is an ancestral land and we owe respect to its owners, so we pay with the tobacco plant," he continued, almost putting forth the clause of a sacred contract between the human newcomers and their forest hosts.

We walked in silence for over 30 minutes along a muddy trail between all kinds of trees coloured in several tones of green. Above, the branches almost embraced one another, while a potpourri of invisible birds momentarily suspended the rhythm of our conversation as we walked through the forest.

At the end of the trail, we found a place with *malocas* (traditional wooden constructions). "And the elder?" the *taita* asked – "keep on walking and you will find his house by the creek," one of the residents replied. After we arrived at the *abuelo's maloca*, we climbed up a few steps to a humble living room where we rested for a while. Fleeting hens and ducks passed by under the wooden house floor, as the *abuela* handed us some *chicha de maíz* (a fermented corn drink).

Early in the morning, we began our bodily preparations for the upcoming *toma* that night. We walked through a forest of *moriche* palms (*Mauritia flexuosa*), tobacco trees, and plantains to a creek about a mile away from the *abuelo's* residence. One of the community leaders was waiting for us with *totumas* ("gourds") filled with the "tea of the Indians." This was how he jokingly named an infusion of the local *yocco* plant (*Paullinia yocco*).

In large quantities, this plant can be used as a purgative, whereas a single cup can evince its tonic properties: "the *yocco* gives strength to work; it purges the body, and it cures the soul," one of the community members said. Light and aligned, our bodies were slowly getting ready to endure the night of the *toma*. As I learned later that day, this ritual of the *ayahuasca* ingestion was particularly important, for it was offered to seal a pact of intercultural exchange and learning between two indigenous communities, the Cofán from the lowlands and the Pastos from the Andean mountains of the Great Cumbal.

An earthly mirror of the cosmos, the *Gran Maloca*, where the *toma* was held, is an ample, sturdy wooden construction with a rich *carana* palm roof (*Mauritia carana*). It is also a boa, a tree-world, and the house of spirits where we ingested the plant after a day-long preparation of bodies and minds. To the right, a forest of tobacco trees moved rhythmically with the wind, while on the left side the *chagra* ("family plot") plants nested, wide open to the sunlight.

"Don't go to the tobacco forest: the spirits live there entangled with the trees and it can be dangerous for you," the *taita* said at some point during the evening. That evening, we started to learn about "where the indigenous law comes from," and what plants can teach us about it. *"De Aquí viene el derecho indígena"* ("indigenous law comes from *Here*") the *taita* said in the midst of the *chuma*. What did he mean by this? I asked myself. Can plants be not only bearers of rights but also law-making entities? And what does such an odd place *"Aquí"* ("Here") represent for the human and other-than-human *people* I met?

Legal plants: How vegetal others invite us to think about environmental law and governance otherwise

Several studies in the fields of Amerindian anthropology and plant science render vegetal beings capable of performing tasks considered exclusive to the human per anthropocentric definitions of language and agency.[12] For instance, plants are deemed intelligent and sentient entities, thus opening up a cascade of theoretical considerations with potential legal interest.[13]

The case of plants invites us to re-think culture (and the law within it) as an exclusive attribute of organisms endowed with brains and nervous systems (Varela, 1999). As plant scientists Viola and Mancuso argue, plants can "breath without having lungs, nourish themselves without having a mouth or stomach, stand erect without having a skeleton, and [...] make decisions without having a brain" (Mancuso and Viola, 2015, p34).

The idea that plants and other organisms are capable of intelligent behaviour expands notions of representation beyond the human and the person as a bounded self. Yet, the modes of representation that plants have are not reducible to those specific to the human; that is, language-based signs (Kohn, 2013). Broadly speaking, I expand the notion of representation usually defined in two complementary ways. First, as the language-mediated outcomes of human perception, that is, the names assigned to exterior objects and the relations between them (Westermann and Mareschal, 2014). And second, as the act of standing for another party by contract or legal right, for example, when humans speak on behalf of others in a congress or a court of law (Vieira and Runciman, 2008). Yet, neither the language-mediated perception of a single and stable reality nor the act of speaking on behalf of others forecloses the possibility of other-than-human (legal) representation and practice beyond symbolic signs.

Anthropologist Eduado Kohn's work with the Runa of the Ecuadorian Amazon offers rich ethnographic evidence on the intrinsic meaning-making and sign-producing capacities of other-than-humans such as plants and animals of interest to my argument. In Amazonia, he argues, life forms create modes of meaning not limited to conventional and language-based signs.

Utilizing Pierce's semiotic framework, Kohn identifies two other representational modalities beyond symbolic signs, namely, icons and indexes. Signs are iconic when they share a likeness with that for which they stand. For example, a picture of your family is an *icon* of your family. Terms such as *splash*, *hiccup*, and *meow* are also iconic.

Indexical signs, on the other hand, express a relation of spatial or temporal continuity with that for which they stand. Or, in other words, an index stands for some physical feature that points to something other than itself, for instance,

> the sound of the palm tree crashing frightened the monkey from her perch [...] The crash, as sign (of danger for the monkey), is not a likeness of the object it represents. Instead, it points to something else [...] this sort of sign (is) an index.
>
> *(Kohn, 2013, p31)*

Thus, what we humans share with non-human species is not only (or even most crucially) our embodiment, Kohn argues, "but the fact that we all live with and through signs [...] signs make what we are" (2013, p9). Non-symbolic representation then is common to all life, so life *is not* without signs. Consequently, signs are not the monopoly of (human) language – let alone legal language – thus living selves represent the world in myriad ways, and this sign-making capacity is intrinsic to all life forms.

Living organisms such as humans, animals, and plants are not only persons, for life recursively exceeds what the law deems as such for the purposes of granting legal rights. Paraphrasing Dewey, *life* "does not signify what the law makes it signify." Quite the opposite, the *law* does signify what life makes it signify. Sign-making life, then, overflows the norm, namely, the classic *ought-to-be* of the law putting "life in order" (Foucault, 1978, p138). The norm, however, cannot exist outside the living where it is embedded (Braverman, 2016), for life constantly exceeds the purview of the law, as well as its person-making machine.

Additionally, if representation is not a property of human persons alone (as "legal fictions"), then representation is not only the human attribution of meaning upon semiotic-devoid matter. Representation, then, becomes a strong property of non-persons; simply put, non-persons-as-relations exist outside the purview of the law-as-a-system-of-norms. A system that governs what kind of life shall be granted with rights.

Paradoxically, however, non-persons are the *rule*, while persons, or what the law defines as such, are the exception (Teubner, 1987). In the example, the *pando* represents the world through other-than-symbolic modalities (indexes and icons), and these representations are essential to its life experience. Marder furthers this argument when asserting that vegetal life expresses itself otherwise and without resorting to vocalization or language. For him,

> aside from communicating their distress when predators are detected in the vicinity by realizing airborne (or in some cases belowground) chemicals, plants, like all living beings, articulate themselves spatially; in a body language free from gestures, they can express themselves only in their postures.
>
> *(Marder, 2013, p75)*

To summarize, as sign-producing entities, non-human selves such as the *pando* operate in a double register. On the one hand, they can be *persons* and thus "signify what the law makes it signify" (Dewey, 1926, p655) through environmental governance models and the rights of nature clause. In this sense, *pando*, as a subject of rights, shall claim legal remedy via the figure of legal guardianship (Stone, 1972).

However, as Roberto Esposito has made clear, it needs to "penetrate the enclosed space of the person" first (2010, p121). That is, the *pando* (and the *pando*'s guardian) needs juridical personality through the codifying legal apparatus, so that it can claim legal standing to be recognized and protected. On the other hand, the *pando* can be a non-person, or the excess of the law. Such non-personality amounts to what escapes the purview of the legal eye, namely, the rhizome that makes up more than half of the *pando*'s biomass beneath the soil (Wohlleben, 2016), making it more-than-a-collection-of-surface-trees.

Insofar as the *pando* is more-than-trees, it occupies a non-personal body. Yet, in order to live, this body does not need to be purified as a person by the legal machine. As a person, however, the *pando* *is* within the purview of an endangered species act, a biological resources protection treaty, or a constitutional clause granting it rights as well.

As a non-person, I wonder if the *pando* would prefer to go unnoticed so that it can occupy the generative excess of the law, that is, the material-semiotic processes that make the law possible through the *pando*'s relational self.

The elusive character of the *pando* self affords an entirely different perspective of the world this vegetal form of life is able to inhabit (Marder, 2013), as well as the kind of legal thinking that emerges from this realization. The law might indeed be an entirely different thing from the perspective of the concealed rhizome, than from the perspective of the visible tree (Kohn, 2018).

Expanding agency to non-persons in the court of justice

The Anthropocene has become a buzzword in today's parlance about climate change, biodiversity loss, and the related socio-ecological impacts of differentiated human agency. Critical legal studies have recently approached the complicated dynamics of multispecies relations in the production of the law, while situating the human within larger assemblages of living and non-living agencies (Grear, 2017; Philippopoulos-Mihalopoulos, 2017; Vermeylen, 2017).

As a system of norms and procedures, the law is almost exclusively depicted as an all-too-human institution. This legal ontology achieves the double task of undermining the material and non-human dimensions of the legal field, while reinforcing a paradigm that separates humans from the larger community of life (Brown and Timmerman, 2015). Does more-than-human life express other kinds of legalities? (Braverman, 2016).

As a positional legal fiction, the notion of personhood conjures certain entities at the expense of others in the legal field. For instance, corporations are considered legal persons, yet persons are not always of the human kind. What is left outside the person-making legal apparatus? This chapter expanded the notion of legal agency beyond the human and the person, while suggesting that plants and forests are not only rights-bearing entities but also sources of legal meaning.

Indigenous practices with plants wield repercussions that extend far beyond local life, and into formal environmental politics in the Andean-Amazonian region. In fact, indigenous relationships with ritual plants exceed ritual spaces to occupy the politics of the everyday, and local dealings with the state law. The ingestion of *Paulina yoccco* and *Banisteriopsis caapi*, for example, plays an active role in decision-making protocols in this region, thereby affording a compelling entry point to other legalities.

Yet, there is a long way before state-oriented legal frameworks incorporate more-than-human vegetal legalities in theory, and practice (Pelizzon and Gagliano, 2015; Braverman, 2018; Davies, 2017). While this emergent jurisprudence may benefit from indigenous cosmologies and plant science alike, symbolic representation distinguishes the law from other fields of practice, and experience. The law is almost exclusively conceived as a lettered practice. Can the plant speak *law*? Does the forest always need the mediation of the botanist, for example, in a court of law?

I drew from Eduardo Kohn's work with the Runa of the Ecuadorian Amazon. "Life thinks," he claims as an ontological premise of his work, thus extending the notion of representation to non-human selves such as animals, plants, and spirits. This chapter expanded the legal framework of agency to engage with the law both as a particular kind of symbolic representation – for example, as a set of positive norms produced by humans – as well as a non-symbolic form of representation vis-à-vis images, processes of materialization, sounds, and experience involving other-than-human selves.

Read as a proposal of legal philosophy, *How Forests Think* (2013) also offers an analytically sophisticated methodology to explore instances of legal meaning such as sonic images of the forest, shamanistic chants, and ecological relations as sources of legal principles and procedures. Such an effort, however, requires a particular mode of attunement to the social worlds that plants and humans co-create through sowing, commensality, and ritual, among other practices. Attempts to expand the notion of the law beyond the symbolic (the legal norm), I consider these practices as sources of legal meaning in their own right.

In fact, the forest's legal agencies might be better expressed in terms of alliance or partnership between interdependent humans, and forests. This move problematizes languages of personhood and legal guardianship that separate humans from the larger community of life. Inter-being alliance thus recognizes the human mind as part and parcel of the larger mind of the forest (Kohn, 2013).

If the human can engage with the forest through science, the human can also learn how to think as/with forests in a court of law. While this methodological question requires independent treatment, appropriate cultural engagement with plants may be one such mode of learning with forests' minds, and the legal protocols they harness for post-extractivist transitions in Amazonia.

While botanists and anthropologists may be summoned to render testimony on behalf of forests and cultures – respectively – it remains an important challenge to explore whether judges should engage with other-than-human beings as they do with humans as expert witnesses.

How far off are we from asking judges and legislators to interact with other-than-humans such as forests and rivers as they decide cases involving these kinds of beings? Should a judge, for example, ingest a ritual plant to understand what the forest *wants* when it comes to a mining license in Amazonia? Could this be considered an appropriate methodology for adjudicating justice in certain cases? Could a judge walk the *páramo* (a fragile ecosystem of the Andes), let it speak the language it speaks best, and then consider this experience a form of witness testimony? Should we elect legislators with proven commitment to listening to forests? How about forest literacy as a requirement for licensing lawyers and adjudicating justice? These are some questions for potential engagement on this thorny issue.

Notes

1 Thanks to Kirsten Anker, Peter G. Brown, Eduardo Kohn, Maria Clara Pardo, Daniel Ruiz, Joshua Sterlin, and the editors of this volume for generous and insightful conversations and comments.
2 In the Latin American context: Protective Action issued by the Provincial Court of Loja, Sentence No. 1121-2011-001, March 30, 2011; Constitution of the Republic of Ecuador 2008, Legislative Decree No. 0, Official Registration 449 October 20, 2008; Colombian Constitutional Court, Sentence C-035 2016; Colombian Constitutional Court, Sentence T-622 2016; Justice Supreme Tribunal STC 4390-2018.
3 Wohlleben writes "plants are perfectly capable of distinguishing their own roots from the roots of other species and even from the roots of related individual" (2016: 17).
4 See Sierra Club vs. Rogers Clark Ballard Morton, Secretary of the Interior et al. 405 U.S. 727 (more) 92 S.Ct 1361, 31 L. Ed. 2nd 636.
5 Living relations (rather than legal persons) produce legal meaning in and of itself, and beyond a purposeful act of personal adjudication. I suggest expanding the notion of agency beyond the human, and beyond the norm.
6 See https://pandopopulus.com/about/pando-the-tree/ (accessed 06.27.2019).
7 This short ethnographic account does not represent the positions of the people I have the opportunity to interact with in Lower Putumayo. I will use different names to protect the identities of peoples and places.
8 Name given to a traditional healer in some parts of the Amazon.
9 The chacruna plant is a perennial shrub used in the preparation of the *ayahuasca* brew. The name comes from the Quechua verb "chaqruy" meaning "to mix." Also, this plant is combined with the *Banisteriopsis caapi* vine for the preparation of the brew also known as yajé. The term comes from the Quechua as well, and it has been translated as the "vine of the soul." For further reference see Daniel Mirante, On the Origins of Ayahuasca. See http://www.ayahuasca.com/ayahuasca-overviews/on-the-origins-ofayahuasca.
10 The *chuma* is the healing vertigo akin to a deep drunken sensation in the context of plant-based rituals in some regions of the Amazon.
11 Thanks to anthropologist Kristina Lyons for sharing this reference.

12 See De la Cadena (2015), Kohn (2013), and Viveiros de Castro (1998). On plant science and anthropological takes on the matter, see Myers (2015). On cognition and
life, see Ingold (2011) and Varela (1999).
13 On plant communication and memory, see Mancuso and Viola (2015) and Trewavas
(2016).

References

Acosta, A., & Martínez, E. (2011). *Naturaleza Con Derechos: De La Filosofía a La Política.*
Quito: Ediciones Abya-Yala.
Anker, K. (2017). "Law as … forest: Eco-logic, stories and spirits in indigenous
jurisprudence." *Law Text Culture, 21*: 191–213.
Atleo, R. (Umeek) (2011). *Principles of Tsawalk: An Indigenous Approach to Global Crisis.*
Vancouver: University of British Columbia.
Berry, T. (1999). *The Great Work. Our Way into the Future.* New York: Bell Tower.
Berry, T., & Swimme, B. (1992). *The Universe Story: From the Primordial Flaring to the
Ecozoic Era—A Celebration of the Unfolding of the Cosmos.* San Francisco, CA: Harper.
Borrows, J. (2016). *Freedom and Indigenous Constitutionalism.* Toronto: University of
Toronto Press.
Braverman, I. (2016). "Introduction: Lively legalities." In I. Braverman (ed.), *Animals,
Biopolitics, Law. Lively Legalities* (pp. 3–18). New York: Routledge.
Braverman, I. (2018). "Law's underdog: A call for more-than-human legalities." *Annual
Review of Law and Social Science, 14*: 127–144.
Brown, P., & Timmerman, P. (2015). *Ecological Economics for the Anthropocene.* New York:
Columbia University Press.
Burdon, P. (2010). "The rights of nature: Reconsidered." *Australian Humanities Review,
49*: 69–89.
Burdon, P. (2012). "A theory of earth jurisprudence." *Australian Journal of Legal Philosophy,
37*: 28–60.
Capra, F., & Mattei, U. (2015). *The Ecology of Law: Towards a Legal System in Tune with
Nature and Community.* Oakland, CA: Berrett-Koehler Publishers.
Cullinan, C. (2011). *Wild Law: A Manifesto for Earth Justice* (2nd ed.). White River
Junction, VT: Chelsea Green Publishing.
Davies, M. (2017). *Law Unlimited: Materialism, Pluralism, and Legal Theory.* London:
Routledge.
De la Cadena, M. (2015). *Earth Beings: Ecologies of Practice.* Durham, NC: Duke University
Press.
Descola, P. (2013). *Beyond Nature and Culture.* Chicago: University of Chicago Press.
Dewey, J. (1926). "The historic background of corporate legal personality." *Yale Law
Journal, 35*(6): 655–673.
Escobar, A. (2016). "Sentipensar con la Tierra: Las Luchas Territoriales y la Dimensión
Ontológica de las Epistemologías del Sur." *Revista de Antropología Iberoamericana, 11*(1):
11–32.
Escobar, A. (2018). *Designs for the Pluriverse: Radical Interdependence, Autonomy, and the
Making of Worlds.* Durham, NC: Duke University Press.
Esposito, R. (2010). "For a philosophy of the impersonal." *CR: The New Centennial
Review, 10*(2): 121–134.
Few, M., & Tortorici, Z. (2013). *Centering Animals in Latin American History.* Durham,
NC: Duke University Press.

Foucault, M. (1978). *The Will to Knowledge: The History of Sexuality*, vol. 1. London: Penguin.

Gagliano, M. (2013). "Persons as plants: Ecopsychology and the return to the dream of nature." *Landscapes: The Journal of the International Centre for Landscape and Language*, 5(2): 1–11.

Gagliano, M., Grimonprez, M., Depczynski, M., & Renton, M. (2017). "Turn in: Plant roots use sound to locate water." *Oecologia*, *184*(1): 151–160.

Gagliano, M., Renton, M., Depczynski, M., & Mancuso, S. (2018). "Experience teaches plants to learn faster and forget slower in environments where it matters." *Oecologia*, *175*: 63–72.

Garver, G. (2013). "The rule of ecological law: The legal complement to degrowth economics." *Sustainability*, *5*: 316–337.

Garver, G. (2019). "A systems-based tool for transitioning to law for a mutually enhancing human–earth relationship." *Ecological Economics*, *157*: 165–174.

Graham, N. (2011). *Lawscape: Property, Environment, Law*. London: Routledge-Cavendish.

Grear, A. (2017). "'Anthropocene, Capitalocene, Chthulucene': Re-encountering environmental law and its 'subject' with Haraway and new materialism." In L. Kotz (ed.), *Environmental Law and Governance for the Anthropocene* (pp. 77–96). Portland, OR: Hart Publishing.

Gudynas, E. (2009). "La Ecología política del giro biocéntrico en la nueva Constitución del Ecuador." *Revista de Estudio Sociales*, *34*: 34–47.

Haraway, D. (2008). *When Species Meet*. Minneapolis, MN: University of Minnesota Press.

Harris, A. P. (2014). "Vulnerability and power in the age of the anthropocene." *Washington and Lee Journal of Energy, Climate, and the Environment*. Research Chapter No. 370.

Ingold, T. (2011). *Being Alive: Essays on Movement, Knowledge, and Description*. New York: Routledge.

Kimmerer, R. W. (2013). *Braiding Sweetgrass: Indigenous Wisdom, Scientific Knowledge and the Teaching of Plants*. Minneapolis, MN: Milkweed Editors.

Kohn, E. (2013). *How Forests Think*. Berkeley, CA: University of California Press.

Kohn, E. (2018). "Forest for the trees: Spirit, psychedelic science, and the politics of ecologizing thought as a planetary ethics." *Living Earth Workshop Chapter*. October, 26–29.

López, E. (2018 [2004]). *Teoría Impura del Derecho. La Transformación de la Cultura jurídica latinoamericana*. Bogota: Legis.

Mancuso, S., & Viola, A. (2015). *Brilliant Green: The Surprising History and Science of Plant Intelligence*. Washington: Island Press.

Marder, M. (2013). *Plant-thinking: A Philosophy of Vegetal Life*. New York: Columbia University Press.

Marder, M. (2016). *Grafts*. Minneapolis, MN: University of Minnesota Press.

Margulis, L., & Sagan, D. (2002). *Acquiring Genomes: A Theory of the Origins of Species*. New York: Basic Books.

Mesa Cuadros, G. (2008). "De la ética del consume a la ética del cuidado: de cómo otro mundo sí es posible desde otra manera de producir y consumir." *Pensamiento Jurídico*, *22*: 333–345.

Myers, N. (2015). "Conversations on plant sensing: Notes from the field." *Nature Culture*, *3*: 35–66.

Pelizzon, A. (2014). "Transitional justice and ecological jurisprudence in the midst of an ever-changing climate." In N. Szablewska & S. D. Bachmann (eds.), *Current Issues in Transitional Justice* (Springer Series in Transitional Justice), vol. 4 (pp. 317–338). Cham: Springer.

Pelizzon, A., & Gagliano, M. (2015). "The sentience of plants: Animal rights and rights of nature intersecting?" *Australian Animal Protection Law Journal, 11*: 5–13.

Philippopoulos-Mihalopoulos, A. (2017). "Critical environmental law in the Anthropocene." In L. Kotzé (ed.), *Environmental Law and Governance for the Anthropocene* (pp. 117–136). Portland, OR: Hart Publishing.

Posner, E. (2014, Dec. 4). "The case against human rights." *The Guardian.* Retrieved from https://www.theguardian.com/news/2014/dec/04/-sp-case-against-human-ri ghts.

Rodríguez, A. (2014). "Así es cómo se empezó a enseñar." *Mundo Amazónico, 5*: 285–295.

Schultes, R. E., & Hofmann, A. (2012). *Plantas de los Dioses.* Mexico City: Fondo de Cultura Económica.

Strathern, M. (2005). *Kinship, Law and the Unexpected: Relatives are Always a Surprise.* Cambridge: Cambridge University Press.

Stephens, P. H. G. (2019). "Why nature experience still exists and matters in the anthropocene." In C. Orr, K. Kish, & B. Jennings (eds.), *Liberty and the Ecological Crisis: Freedom on a Finite Planet* (pp. 128–141). Routledge.

Stone, C. (1972). "Should trees have standing? Toward legal rights for natural object." *Southern California Law Review, 45*: 450–501.

Teubner, G. (1987). *Juridification of Social Spheres: A Comparative Analysis in the Areas of Labor, Corporate, Antitrust and Social Welfare Law.* Berlin: W. de Gruyter.

Trewavas, A. (2016). "Plant intelligence: An overview." *BioScience, 66*(7): 542–551.

Varela, F. (1999). *Ethical Know-How: Action, Wisdom and Cognition.* Stanford, CA: Stanford University Press.

Vargas Roncancio, I., Temper, L., Sterlin, J., Smolyar, N. L., Sellers, S., Moore, M., … Babcock, M. (2019). "From the Anthropocene to mutual thriving. An agenda for higher education in the Ecozoic." *Sustainability, 11*(12): 3312.

Vermeylen, S. (2017). "Materiality and the ontological turn in the Anthropocene: Establishing a dialogue between law, anthropology, and eco-philosophy." In L. Kotzé (ed.), *Environmental Law and Governance for the Anthropocene* (pp. 137–162). Portland, OR: Hart Publishing.

Vieira, M. B., & Runciman, D. (2008). *Representation.* Cambridge: Polity Press.

Vivieros de Castro, E. (1998). "Cosmological diexis and Amerindian perspectivism." *Journal of the Royal Anthropological Institute, 4*(3): 469–489.

Viveiros de Castro, E. (2014). *Cannibal Metaphysics.* Minneapolis, MN: Univocal.

Westermann, G., & Mareschal, D. (2014). "From perceptual to language-mediated categorization." *Philosophical Transactions of the Royal Society of London: Series B, Biological Sciences, 369*: 1634.

Winter, S. (2001). *A Clearing in the Forest. Law, Life, and Mind.* Chicago: University of Chicago Press.

Wolfe, C. (2016). "Foreword: 'Life' and 'the living,' law and norm." In I. Braverman (ed.), *Animals, Biopolitics, Law. Lively Legalities* (p. xiii). New York: Routledge.

Wohlleben, P. (2016). *The Hidden Life of Trees: What They Feel, How the Communicate— Discoveries from a Secret World.* Vancouver: Greystone Books.

Youatt, R. (2017). "Personhood and the rights of nature: The new subjects of contemporary earth politics." *International Political Sociology, 11*: 39–54.

17

FROM THE ECOLOGICAL CRISIS OF THE ANTHROPOCENE TO HARMONY IN THE ECOZOIC

Christopher J. Orr and Peter G. Brown

A heroic voyage of thought and action

There is a deeply unsettling mismatch between the human systems that modern societies have constructed and the natural processes that enable and sustain those societies. The global ecological crisis we term the Anthropocene presents an unprecedented situation that threatens the future of humans and other life forms as well as the habitability of Earth itself. Through numerous actions such as carbon emissions, species extinction, soil erosion, and nutrient pollution, humans are overwhelming the Earth's systems, disrupting and degrading the capacities of these systems to absorb wastes, regenerate themselves, and support life (Steffen et al. 2015). These pressures will only be compounded by rapidly rising sea levels, large-scale forced immigration of humans and other species, desertification, and increasing frequency and severity of extreme weather events. Undermining and destabilizing these ecological foundations creates hard trade-offs and wicked dilemmas between human and non-human well-being (O'Neill et al. 2018; Kish and Quilley 2017).

This mismatch cannot simply be addressed through the same thinking that brought about the problem in the first place, but requires an entirely new relationship with nature. Although some speak of the Anthropocene as something that can be good or at least have good seeds (Asafu-Adjaye et al. 2015; Bennett et al. 2016), we take strong exception to this. If the Anthropocene is defined as a situation in which "the human imprint on the global environment is now so large that the Earth has entered a new geological epoch" so that humans are a "global geophysical force" in terms of Earth System functioning (Steffen et al. 2011, p741), the human–Earth relationship that characterizes this situation is in no way desirable or good. Indeed, it is hard to see how mass extinction of species and cultures or global sea level rise are in any way good. As such, there are both moral and practical imperatives to reject and abandon this relationship with nature and to exit the Anthropocene as soon as possible to the greatest degree possible.

Not only do we need to change our dominant ideas and values, but also the institutions and practices that legitimate and sustain them. This implicates the entire complex of ideas, institutions, and ways of life that embody this dominating and destructive relationship with nature. We might think of the current period as what Karl Jaspers called an axial age; a pivotal age in which old certainties lose their validity, while new concepts and ways of knowing and being appear. Dale Jamieson provides a sobering but perhaps realistic view of life in the Anthropocene:

> We will have to manage and live as best we can, and hope that the darkest scenarios do not come to pass. We will have to abandon the Promethean dream of a certain, decisive solution and instead engage with the messy world of temporary victories and local solutions while a new world comes into focus.
>
> *(Jamieson 2014, p15)*

Cultural values and visions of the good life play a pivotal role in weathering such transitions. As Dale Jamieson and Marcello di Paola point out,

> A political theory for the Anthropocene [...] will have to coevolve with it while at the same time continuing to indicate normatively acceptable directions for much-less-than-ideal politics. To preserve liberal democratic values in the Anthropocene, such theory may need to call into question some of liberal democracy's own basic constructions.
>
> *(Jamieson and di Paola, 2016, p278)*

Rather than rejecting liberty and agency out of hand, these concepts will be central in dealing with the ecological crisis of the Anthropocene. On the one hand, there is a need for a radical departure from dominant conceptions of liberty and agency. In the Anthropocene, with the carbon sink full, every action has impacts on humans and non-humans alike. John Locke's liberal maxim that one is free to do what one wants "at least where there is enough, and as good, left in common for others" (Locke 1884, p116) implies that many, if not most, current actions have negative moral consequences. There are few, if any, neutral acts. This is not a small problem because it undermines the foundational assumptions of economic and political liberalism. Hence, the Anthropocene reduces society's options, curtailing human agency and freedom to make meaningful and morally responsible choices. Moreover, unbridled freedom in the pursuit of self-interest abstracted from the social and material world legitimates the overarching thrust of this ecologically destructive cultural complex and both disregards and rejects any ethical relationship of responsibility, humility towards, or care for nature (Ophuls 1977). On the other hand, these notions also represent the ontological and ethical core of the dominant cultural complex, thus providing an important fulcrum in this transformation to a new human–Earth relationship.

This chapter calls for a heroic voyage of thought and action to bring into being a new human–Earth relationship grounded in a positive vision of the Ecozoic and tempered by educated hope. It argues that regrounding the human–Earth relationship entails a foundational shift in the underlying ontology of modernity. It shows how regrounding the human–Earth relationship through such a seismic shift has important implications for both agency and liberty. Next, it develops Thomas Berry's vision of the Ecozoic, characterized by a relationship of resonance with nature and shows how liberty and agency play important roles in this vision. It concludes with a call for educated hope for the Ecozoic, a possibility still visible on the horizon.

Regrounding the human–Earth relationship and the myth of freedom

The Anthropocene demands not only compassionate retreat (Brown and Schmidt 2010), but also a regrounded relationship between humans and nature. We cannot only retreat and leave space for nature because simply withdrawing relies on the same thinking and relationship of separation from nature that epitomizes the Anthropocene. Moving beyond human–nature dualisms, mere human presence is not inherently bad or unnatural; rather, transformation of nature into artefacts through instrumentalization, and human imposition lead to increasing abstraction and domination. The implication is that, more than simply leaving space for nature, a truly ecological civilization would be in constant conversation and relation with the non-human world. Thus, as Cannavo (this volume) emphasizes, the challenge of the human–Earth relationship is to find a balance between arrogant humanism and ecological quietism.

Regrounding the human–Earth relationship is not primarily epistemic, but ontological, implicating a shift in our fundamental assumptions, cultural myths, and practices so as to bring another world into being. The modern sciences teach us to "think of the universe as a collection of objects rather than as a communion of subjects" (Berry 1999, p16). They teach us to think humans are separate from nature, and that our relationship is one of domination, control, domestication, and development of "dead" inanimate nature. As a result, we attempt to "domesticate" wild nature, replacing the creative patterns of the universe with our own (Berry 1999, pp50–51).

The dominant human–Earth relationship that has led to the global ecological crisis called the Anthropocene is characterized by a number of ontological presuppositions. Assumptions of this dominant belief system that have been disproven include: (1) reductionism that results in a mechanistic worldview of a clockwork universe; (2) the ideas of control and complete knowledge in the face of uncertainty; (3) the myth of the rational, self-interested individual; (4) the radical separation of mind from body of Cartesian dualism; (5) social atomism rejecting both social relations and ecological embeddedness; (6) the dominance and ownership of nature as property or resources for human profit and use; and (7) the equation of prosperity with material progress and economic growth (Callicott 2015; Lakoff 2008; Levine et al. 2015; Merchant 2003; Purdey 2010). Indeed, nearly all of these core tenets have been undermined by contemporary science that recognizes that humans are embedded in and are part of an evolving universe.

Some suggest that the entire system of beliefs that undergirds modernity is deeply flawed, based on myths that are at odds with reality. While this system of beliefs pervades modern society and benefits from wide legitimacy, it contains deep contradictions that call into question its social and environmental merits (Purdey 2010). These myths, treated as common sense, are embedded in the dominant institutions and practices of modern life, and are part of the way we shape and inhabit the world. The result is an increasingly human-centric, self-referential world: "we make everything referent to the human as the ultimate source of meaning and value, although this way of thinking has led to catastrophe for ourselves as well as a multitude of other beings" (Berry 1999, p18).

If we are to leave space for nature to be without overwhelming human influence and control, this also requires us to stop reordering the non-human world according to our increasingly isolated, disembedded and self-referential experience. And yet, the isolated and self-referential experience of the cell phone- and fossil fuel-addicted, nine-to-five commuter culture of suburban living, has constructed and contributes to maintaining the myth of freedom. From a commonsense point of view, it might appear that we are free to do as we wish in the Western democracies. As citizens, we can travel as we wish, eat what we want, attend "good" universities, and marry and divorce as desired. But if you are middle class and are not traveling to some (formerly) exotic place then there must be something wrong with you. William Buckley said if you mention religion more than once at a dinner party you will never be invited back. We might update this observation by noting that if you mention that we live in a carbon-constrained world you risk a similar fate. A plain fact is that we live in an age of exogenously constructed desires. The power and sophistication of marketing and manipulation have grown stronger and more and more pervasive, while the internet, once hailed as an advance in sharing information, is now used to kindle and stimulate desires. We are not only slaves to our own desires, but to the desires of others about our desires. As Thoreau noted, "men have become the tools of their tools" (2004 [1854], p29). The market now manufactures the person.

This myth of human freedom is built upon and presupposes a conception of humans as agents abstracted from others and the world. The Enlightenment resulted in the demystification of the world, the separation of humans from nature, mind from body, and the objectification of "dead" nature (Hornborg 2016). This gave rise to the distinction between humans as subjects or agents as opposed to non-humans as objects to be controlled, shaped, and used according to human desires. A human–Earth relationship of domination and exploitation assumes this notion of humans as separate from others and from nature, rather than dependent upon and enabled by nature. Likewise, this notion of human agency legitimates and makes possible human freedom to dominate, use, and exploit nature without care and respect.

Yet, the Anthropocene epitomizes a world in which human attempts to dominate nature reveal the limits to human agency and undermine our ability to make meaningful choices, while at the same time bringing forth an eruption of non-human forces: "The contradiction in our relation to nature is that the more vigorously we attempt to force its agreement with our own designs the more subject we are to its indifference, the more vulnerable to its unseeing forces" (Carse 1986, p144). In other words, the Anthropocene remains a paradox of agency only as insofar as we continue to attempt to dominate nature: "The more power we exercise over natural processes the more powerless we become before it" (Carse 1986, p144).

Rather than seeing humans as autonomous individuals separate from nature, contemporary science has emphasized human and non-human interdependence in meaningful processes of being and becoming. Human agency is necessarily relational and dependent upon non-human entities and processes, even while subjectivity, agency, and semiosis are inherent to the universe. The world is full of signs and signals, from the interactions between eagles and fish and insects, to forests of trees and other beings that transmit information and share nutrients (Kohn 2013). This ontological position recognizes that non-human entities and processes are not necessarily inert, meaningless, and controllable, but characterized by semiosis (Kohn 2013; Rice 2013). Moreover, this undercuts both the ontological separation of and differentiation between humans and non-humans, respectively, as subjects and objects. If we are to ascribe agency to humans, something similar applies to different degrees and in different ways to non-human beings (Rice 2013). This shift in underlying ontology has profound implications for both how we envision the human–Earth relationship and the role of liberty.

At home in the Ecozoic: A relationship of harmony and resonance with nature

A positive vision is a creative practice that shows us what a desirable world could be. Envisioning suspends disbelief, opening up opportunities that would otherwise not be considered or pursued. In turn, this can create the social preconditions for pathways that would otherwise not be available. In this way, a positive vision is a way of expanding our options and exercising our agency. Visioning itself can be a prefigurative act. In other words, positive visions keep our eye on the horizon of reality, engaging with concrete utopias as we – along with the rest of nature – bring into being another world.

Thomas Berry's Ecozoic provides a positive vision that contrasts starkly with the Anthropocene and provides an anchor that might pivot us towards a new relationship with the Earth. Berry envisions the Ecozoic as a world in which "humans would be present to the planet in a mutually beneficial manner" (1999, p3). The key to a mutually enhancing human–Earth relationship is an ethic of respect, care, and reciprocity. Thus, "mutually enhancing" implies a relationship in which humans contribute to the flourishing and regeneration of the Earth. There are numerous anthropological examples of peoples who have lived for thousands of years on the Earth without ecological overshoot and collapse (Brody 2001; Mann 2006), providing hope that such regenerative relationships are indeed possible.

If the Anthropocene emphasizes human limits to knowledge, control, and ability to change the course of the Earth system, the Ecozoic accepts the world as it is and attempts to find ways humanity can be at home in the universe (Kauffman 1995). Such a relationship is grounded in recognition of humans and non-humans as subjects of the Earth community. Humans are part of the creative process of becoming of the universe, but are merely members of a community of subjects in that process. Moreover, human creativity flourishes only inasmuch as we are members of that community of subjects.

This process of being and becoming part of the universe can be understood as a relationship of resonance. A relationship of resonance is both a state of being and an ongoing process in which human actions leave space for non-human nature, its ways of being, and the creative patterns of the universe. A relationship of resonance rejoices in existing patterns and attempts to amplify and elaborate on their beauty. This requires attentive listening and for the listener to be actively engaged with the music to stay in tune and in harmony. Attention to natural patterns can help inform human–Earth relationships through diverse facets of our lives, from architecture and the built environment (Alexander 1977) to our systems of governance (Ophuls 2013), to ecological education and ethics (Leopold 1966). A relationship of resonance is engaged and responsive, aligned with the patterns of nature. Once we accept and learn to be at home in the universe, discomfort with the Anthropocene turns to wonder and appreciation in the Ecozoic. Coming to view ourselves as part of the cosmos brings awe and joy, but also the capacity to celebrate its wonders (Berry 1999).

This contrasts starkly with the dissonance of human dominance and control. Human intrusion and imposition, when pushed too far, interfere with nature's existing tendencies and prevent its diverse and often self-regulating capacities from being expressed. For example, attentive agricultural practices may help rebuild soil and sustain biodiversity, while industrial agricultural practices solely focused on yield typically contribute to immense soil erosion and species loss. When taken too far in any one direction or dimension, human amplification of natural tendencies – such as agricultural production – can be disruptive and harmful.

Importantly, amplification and feedback can lead to increasing abstraction and disembeddedness so that the amplified noise overwhelms all other sounds.

Like a speaker placed too close to a microphone, society will only hear its own high-pitched screeching. Something like this has occurred with the increasing monetization and financialization of the global economy. A high-tech, media-immersed, urban lifestyle is the antithesis of a relation of harmony and resonance. Although many modern beliefs may have partial basis in reality, many of them have become untethered from this grounding in an increasingly self-referential and abstracted world. Avoiding such abstraction requires a close, intimate relationship with nature, or the listener will forget who they are accompanying. Human ways of being must complement non-human patterns; not overwhelm them. While there are many songs that might be sung, only when human cultures find ways to resonate in harmony with nature – and each other – can they form an enduring whole.

Such a relationship represents a process of becoming in harmony with the Earth. This requires attention, listening, and engagement with nature through diverse practices, such as science, indigenous ways of knowing and being, diverse cultural traditions and ways of being, and nature experience. Through these practices, we might engage in ecological re-civilizing processes that attune us and bring us more in resonance with natural patterns and ways of being. These practices allow us to rejoice in the creative beauty of the symphony of the Earth and add our part to it. Play in the Ecozoic is a process of being and becoming that resonates in harmony with the chorus of the Earth and its rhythms.

What would the world look like after the death of liberal individualism? Liberty and agency are central to bringing the Ecozoic into being, but not in any modern way they are conceived. If the Ecozoic entails an ontological shift towards recognition of non-human others as equal members of a community, the ethical kernel of liberty is not discarded, but enhanced and deepened through this relationship of resonance. What is this kernel of liberty? Ecological liberty not only rejects domination, coercion, and exploitation, but also rests on our ability to flourish in being and becoming at home in the universe. We might suggest that Macpherson's (1973) counterextractive liberty applies not only to humans, but also, quite literally, to the extractive processes of human domination of nature. In contrast to domination, recognition of non-human others does not narrow our view of freedom, but expands it to encompass our relationships and potentialities in relating and responding to nature and its potential to flourish. In the Ecozoic, human freedom is not only relational but also a way of being that recognizes and allows space for non-human being and becoming: "Human freedom is not a freedom over nature; it is the freedom to be natural, that is, to answer to the spontaneity of nature with our own spontaneity" (Carse 1986, p144). And the more deeply our culture respects nature and allows space for its creativity, surprise, and unpredictability, "the more our culture will embody a freedom to embrace surprise and unpredictability" (Carse 1986, p144). As such, a relationship of resonance means that human freedom and, more broadly, ecological liberty alike are mutually enabled and enhanced.

From the emergency of the Anthropocene to educated hope for the Ecozoic

The global ecological crisis we call the Anthropocene presents an "earth-shattering" new situation (Hamilton et al. 2015). Merely coming to terms with the enormity and implications of this new context should give us pause. Reason is called for in this dark time (Jamieson 2014). Even more sobering is one scenario, as Sartre prefigured in *No Exit*, that there is no escape from the ecological crises of the Anthropocene. Indeed, important figures in the debate think that climate change has or is likely to become unstoppable – with massive methane releases from the previously frozen tundra likely a major factor – and that Earth will become too inhospitable to support complex life as we know it (Wallace-Wells 2019). Some argue that in the face of this crisis conceptual critiques are like rearranging the deck chairs on the Titanic: what is the point if our fundamental ways of life will look so dramatically different that contemporary concepts may be nonsensical? (Mulgan 2011).

We argue that there is also room for educated hope. In *The Principle of Hope*, Ernst Bloch distinguished between abstract and concrete utopia. While abstract utopia is mere aspiration without realistic possibility, concrete utopia aspires to bring into being one of a range of possible realities. In a sense, concrete utopia is both realistic and aspirational: "Concrete utopia stands on the horizon of every reality" (Bloch 1986, p223). Educated hope is borne out of, and articulates the relationship between, aspiration and real possibility and attempts to bring into being concrete utopias glimpsed at the horizon of reality. It accounts for the possible realities on the horizon and translates wishful thinking into wilful and effective action.

The Ecozoic is one possible reality that remains visible and perhaps even viable on the horizon. Being at home in the Ecozoic does not mean naive optimism for a return to the tranquillity of the Holocene. The coming decades may be along the road from the Anthropocene to the Ecozoic; or they may be a litany of ecological crises. The processes involved will play out of over days, decades, centuries, millennia, and deep geological time. How we conceive of and enact liberty and agency may help determine if we can walk the path to the Ecozoic, are among humanity's final Earth-bound voyagers, or fall somewhere in between. A relationship of resonance with the non-human world can help guide us in responding with freedom and creativity to the spontaneity and natural rhythms of the Earth. In exploring our harmonies and resonances with and as part of nature, we may yet find ways of bringing the Ecozoic into being.

How does this endeavour contribute to bringing the Ecozoic into being? As we move deeper into the Anthropocene, the global ecological crisis heralds a period of intense experimentation in political theory and practice (Jamieson and di Paola 2016) that can only be enriched through interrogation of and deep reflection on the core values and presuppositions of modernity such as liberty and agency. Indeed, in the Anthropocene, our concepts, values, and political institutions may look nothing like our current ones. Despite these changes, in the current climate of populism and impending ecological crisis it is now more important than ever to preserve, and also to nourish the ethical core of these concepts. As such, these arguments do not end with futility. Rather, they open constructive dialogues, identify hurdles, and illuminate unforeseen possibilities and prefigurative strategies. Instead of fostering resignation at the paltry prospects for meaningful action in the face of the enormity of the implied changes (Stoner and Melathopoulos 2015), they show how reconsidering liberty in the Anthropocene can engender educated hope by expanding individual and collective agency, and the potential to bring into being another world. Therefore, deep reflection on our current situation is essential to truly educated hope, because it helps us understand and bring into being the possible realities on the horizon.

Bibliography

Alexander, C. (1977). *A Pattern Language: Towns, Buildings, Construction*. New York, NY: Oxford University Press.

Asafu-Adjaye, J., Blomqvist, L., Brand, S., Brook, B., DeFries, R., Ellis, E. C., ... Teague, P. (2015). *An Ecomodernist Manifesto*. Retrieved from http://www.ecomodernism.org/manifesto-english/

Bennett, E. M., Solan, M., Biggs, R., McPhearson, T., Norström, A.V., ... and Xu, J. (2016). "Bright spots: Seeds of a good Anthropocene." *Frontiers in Ecology and the Environment, 14*(8): 441–448.

Berry, T. (1999). *The Great Work: Our Way into the Future*. New York: Three Rivers Press.

Bloch, E. (1986). *The Principle of Hope*. Cambridge, MA: MIT Press.

Brody, H. (2001). *The Other Side of Eden: Hunters, Farmers, and the Shaping of the World*. New York, NY: North Point Press.

Brown, P., & Schmidt, J. (2010). *Water Ethics: Foundational Readings for Students and Professionals*. Washington, DC: Island Press.

Callicott, J. B. (2015). "Science as myth (whether sacred or not), science as prism." *Journal for the Study of Religion, Nature and Culture, 9*(2): 154–168. doi:10.1558/jsrnc. v9i2.27264.

Carse, J. P. (1986). *Finite and Infinite Games*. New York: Ballantine.

Hamilton, C., Gemenne, F., & Bonneuil, C. (2015). *The Anthropocene and the Global Environmental Crisis: Rethinking Modernity in a New Epoch*. Florence, KY, USA: Routledge.

Hornborg, A. (2016). *Global Magic: Technologies of Appropriation from Ancient Rome to Wall Street*. New York, NY: Springer.

Jamieson, D. W., & Di Paola, M. (2016). Political theory for the Anthropocene. In Held, D. and Maffettone, P. (eds) *Global Political Theory*, 254–280. Oxford: Polity Press.

Jamieson, D. (2014). *Reason in a Dark Time: Why the Struggle against Climate Change Failed—and What It means for our Future*. Oxford; New York: Oxford University Press.

Kauffman, S. A. (1995). *At Home in the Universe: The Search for Laws of Self-organization and Complexity*. Oxford, New York: Oxford University Press.

Kish, K., & Quilley, S. (2017). "Wicked Dilemmas of scale and complexity in the politics of degrowth." *Ecological Economics, 142*: 306–317.

Klein, N. (2014). *This Changes Everything: Capitalism vs. the Climate*. Toronto, ON: Vintage Canada.

Kohn, E. (2013). *How Forests Think: Toward an Anthropology beyond the Human*. Berkeley: University of California Press.

Lakoff, G. (2008). *The Political Mind: Why You Can't Understand 21st-Century Politics with an 18th-Century Brain*. New York: Viking.

Leopold, A. (1966). *A Sand County Almanac: With Essays on Conservation from Round River*. New York: Ballantine Books.

Levine, J., Chan, K. M., & Satterfield, T. (2015). "From rational actor to efficient complexity manager: Exorcising the ghost of Homo economicus with a unified synthesis of cognition research." *Ecological Economics, 114*: 22–32.

Locke, J. (1884). *Two Treatises on Civil Government*. London: George Routledge and Sons.

Macpherson, C. B. (1973). *Democratic Theory: Essays in Retrieval*. Oxford, UK: Oxford University Press.

Mann, C. C. (2006). *1491: New Revelations of the Americas before Columbus*. New York: Vintage Books.

Merchant, C. (2003). *Reinventing Eden: The Fate of Nature in Western Culture*. New York: Routledge.

Mulgan, T. (2011). *Ethics for a Broken World: Imagining Philosophy after Catastrophe*. Montreal: McGill-Queen's University Press.

O'Neill, D. W., Fanning, A. L., Lamb, W. F., & Steinberger, J. K. (2018). "A good life for all within planetary boundaries." *Nature Sustainability, 1*(2): 88.

Ophuls, W. (1977). *Ecology and the Politics of Scarcity: Prologue to a Political Theory of the Steady State*. San Francisco: W.H. Freeman.

Ophuls, W. (2013). *Sane Polity: A Pattern Language*. North Charleston, SC: CreateSpace.

Purdey, S. J. (2010). *Economic Growth, the Environment and International Relations: The Growth Paradigm*. London, New York: Routledge.

Rice, J. (2013). "Further beyond the Durkheimian problematic: Environmental sociology and the co-construction of the social and the natural." *Sociological Forum, 28*: 236–260. Wiley Online Library.

Steffen, W., Richardson, K., Rockstrom, J., Cornell, S. E., Fetzer, I., Bennett, E. M., … Sorlin, S. (2015). "Planetary boundaries: Guiding human development on a changing planet." *Science, 347*(6223): 736–746.

Steffen, W., Grinevald, J., Crutzen, P., & McNeill, J. (2011). "The anthropocene: Conceptual and historical perspectives." *Philosophical Transactions of the Royal Society A: Mathematical, Physical and Engineering Sciences, 369*(1938): 842–867.

Stoner, A., & Melathopoulos, A. (2015). *Freedom in the Anthropocene: Twentieth-century Helplessness in the Face of Climate Change*. New York, NY: Palgrave Macmillan.

Thoreau, H. D. (2004). *Walden*. Princeton, NJ: Princeton University Press.

Wallace-Wells, D. (2019). *The Uninhabitable Earth: Life after Warming*. New York, NY: Crown/Archetype.

INDEX

Page numbers in **bold** indicate tables. Page numbers in *italics* indicate figures.